Statistical Analysis Techniques in Particle Physics

Statistical Analysis Techniques in Particle Physics

Contributors

Denis Perret-Gallix et al.

AURIS
Reference

www.aurisreference.com

Statistical Analysis Techniques in Particle Physics

Contributors: Denis Perret-Gallix et al.

Published by Auris Reference Limited

www.aurisreference.com

United Kingdom

Statistical Analysis Techniques in Particle Physics

ISBN: 978-1-78154-887-5

British Library Cataloguing in Publication Data
A CIP record for this book is available from the British Library

Printed in the United Kingdom

Exclusively distributed by CBS Publishers & Distributors Pvt. Ltd.

Sales & Distribution Rights only for India, Pakistan, Bangladesh, Sri Lanka, Nepal and Bhutan. This book is not to be sold outside these territories.

Contents

List of Abbreviations

BSM	Beyond Standard Model
LHC	Large Hadron Collider
NLO	Next-to-Leading Order
NLO	Next-to-Leading Order
PDF	Parton Distribution Function
TOE	Theory of Everything
QFT	Quantum Field Theory
QED	Quantum Electrodynamics
FDG	Fluoro-Deoxy-Glucose
PET	Positron Emission Tomography
TBF	Three-Body Forces
BHFA	Brueckner-Hartree-Fock Approximation
BHF	Brueckner-Hartree-Fock
GRB	Gamma Ray Burst
HTR-M	High Temperature Gas Cooled Reactor
SCM	Stiffness Confinement Method
BAEC	Bangladesh Atomic Energy Commission
HPRWMU	Health Physics and Radioactive Waste management Unit
HPGe	High Purity Germanium
MQW	Multiple-Quantum-Well
HTR-M	High Temperature Gas Cooled Reactor
SCM	Stiffness Confinement Method

List of Contributors

Denis Perret-Gallix
LAPP (IN2P3/CNRS), France

Hsuan Tung Peng
Institute of Atomic and Molecular Sciences, Academia Sinica, Taipei 10617, Taiwan
Department of Physics, National Taiwan University, Taipei 10617, Taiwan

Yew Kam Ho
Institute of Atomic and Molecular Sciences, Academia Sinica, Taipei 10617, Taiwan

Hans W. Giertz
GiertzTech AB, Gnesta, Sweden

Z. J. Ajaltouni
Laboratoire de Physique Corpusculaire de Clermont-Ferrand IN2P3/CNRS
Université Blaise Pascal F-63177, Aubière Cedex, France;

Edwin Zong
Medicine Department Oasis Medical Group, Bakersfield, USA

Edwin Zong
Oasis Medical Group Inc., Bakersfield Ca USA

H. M. M. Mansour
Physics Department, Faculty of Science, Cairo University, Egypt

Edwin Zong
Medicine Department, Oasis Medical Group Inc., Bakersfield, California, USA

H. Mansour
Physics Department, Faculty of Science, Cairo University, Giza, Egypt

Hend M. Saad
Nuclear and Radiological Regulatory Authority, Ahmed Al Zomor St., Nasr City, Cairo, Egypt

M. Aziz
Nuclear and Radiological Regulatory Authority, Ahmed Al Zomor St., Nasr City, Cairo, Egypt

M. M. Ahmed
Department of Physics, Jagannath University, Dhaka, Bangladesh

S. K. Das
Department of Physics, Jagannath University, Dhaka, Bangladesh

M. A. Haydar
Health Physics & Radioactive Waste Management Unit, INST, BAEC, Savar, Dhaka, Bangladesh

M. M. H. Bhuiyan
Health Physics & Radioactive Waste Management Unit, INST, BAEC, Savar, Dhaka, Bangladesh

M. I. Ali
Health Physics & Radioactive Waste Management Unit, INST, BAEC, Savar, Dhaka, Bangladesh

D. Paul
Health Physics & Radioactive Waste Management Unit, INST, BAEC, Savar, Dhaka, Bangladesh

Mohamed Y. Abou-zeid
Department of Mathematics, Faculty of Science, Tabuk University, Tabuk, KSA
Department of Mathematics, Faculty of Education, Ain Shams University, Roxy, Cairo, Egypt

Seham S. El-zahrani
Department of Mathematics, Faculty of Science, Tabuk University, Tabuk, KSA

Hesham M. Mansour
Department of Physics, Faculty of Science, Cairo University Giza, Egypt

B. M. R. Faisal
Department of Environmental Sciences, Jahangirnagar University, Dhaka-1342, Bangladesh

M. A. Haydar
Health Physics and Radioactive Waste Management Unit (HPRWMU), Institute of Nuclear Science and Technology (INST), Atomic Energy Research Establishment (AERE), Bangladesh Atomic Energy Commission (BAEC), Savar, Dhaka-1349, Bangladesh

M. I. Ali
Health Physics and Radioactive Waste Management Unit (HPRWMU), Institute of Nuclear Science and Technology (INST), Atomic Energy Research Establishment (AERE), Bangladesh Atomic Energy Commission (BAEC), Savar, Dhaka-1349, Bangladesh

D. Paul
Health Physics and Radioactive Waste Management Unit (HPRWMU), Institute of Nuclear Science and Technology (INST), Atomic Energy Research Establishment (AERE), Bangladesh Atomic Energy Commission (BAEC), Savar, Dhaka-1349, Bangladesh

R. K. Majumder
Nuclear Minerals Unit, Atomic Energy Research Establishment (AERE), Bangladesh Atomic Energy

M. J. Uddin
Department of Environmental Sciences, Jahangirnagar University, Dhaka-1342, Bangladesh

H. M. Mansour
Department of Physics, Faculty of Science, Cairo University, Giza, 12613, Egypt

A. A. I. Khalil
National Institute of Laser Enhanced Sciences, (NILES), Cairo University, Giza, 12613, Egypt

M. Y. Helali
National Research Institute of Astronomy and Geophysics, Helwan, Cairo, Egypt

M. Mansour
National Institute of Laser Enhanced Sciences, (NILES), Cairo University, Giza, 12613, Egypt

H. Mansour
Physics Department, Faculty of Science, Cairo University, Giza, Egypt

Hend M. Saad
Nuclear and Radiological Regulatory Authority, Ahmed Al Zomor St., Nasr City, Cairo, Egypt

M. Aziz
Nuclear and Radiological Regulatory Authority, Ahmed Al Zomor St., Nasr City, Cairo, Egypt

Preface

Modern analysis of HEP data needs advanced statistical tools to separate signal from background. Statistical Analysis Techniques in Particle will be of interest to almost every high energy physicist, and, due to its coverage, suitable for students. In first chapter, after a short introduction to particle physics and to the related theoretical framework, we will review some of the computing techniques that have been developed to make these calculations automatic. In second chapter, we study the correlation of the ground state of an N-particle Moshinsky model by computing the Shannon entropy in both position and momentum spaces. In third chapter, all energy in the universe, here called energy quanta, originates from a singularity at the centre of the universe. The electron and the atom are completely passive; they absorb energy quanta by means of forced damped oscillators and the absorbed energy quanta are then re-emitted. The aim of fourth chapter is to highlight the common aspects between Symmetry in Physics and the Relativity Theory, particularly Special Relativity. After a brief historical introduction, emphasis is put on the physical foundations of Relativity Theory and its essential role in the clarification of many issues related to fundamental symmetries. Fifth chapter proposes on the origin of particles and the objective of sixth chapter is to uncover the common origin for both mass and force in the universe along with their evolutionary path in the discipline of Newtonian physics. In seventh chapter, we extend the calculation to present the single particle potentials for the proton and neutron using modern nucleon-nucleon (NN) potentials in the framework of (BHFA). The primary objective of eighth chapter is to uncover the real mechanic of dark matter, big bang and evolution process of the universe in the discipline of basic physics. In ninth chapter, the point kinetics equations are solved numerically using the stiffness confinement method (SCM). The aim of eleventh chapter is to investigate the effect of mixed convection heat and mass transfer on pulsatile flow of a non-Newtonian fluid which is obeying the rheological equation of state due to Ree-Eyring's stress-strain relation. Twelfth chapter presents an assessment of natural radioactivity and associated radiation hazards in topsoil of savar industrial area, Dhaka, Bangladesh. Fourteenth chapter describes the techniques used by Particle Physicists for dealing with statistical problems, and also some of the open statistical questions.

Chapter 1

COMPUTATIONAL PARTICLE PHYSICS FOR EVENT GENERATORS AND DATA ANALYSIS

Denis Perret-Gallix

LAPP (IN2P3/CNRS), France

ABSTRACT

High-energy physics data analysis relies heavily on the comparison between experimental and simulated data as stressed lately by the Higgs search at LHC and the recent identification of a Higgs-like new boson. The first link in the full simulation chain is the event generation both for background and for expected signals. Nowadays event generators are based on the automatic computation of matrix element or amplitude for each process of interest.

Moreover, recent analysis techniques based on the matrix element likelihood method assign probabilities for every event to belong to any of a given set of possible processes. This method originally used for the top mass measurement, although computing intensive, has shown its efficiency at LHC to extract the new boson signal from the background.

Serving both needs, the automatic calculation of matrix element is therefore more than ever of prime importance for particle physics. Initiated in the 80's, the techniques have matured for the lowest order calculations (tree-level), but become complex and CPU time consuming when higher order calculations involving loop diagrams are necessary like for QCD processes at LHC. New calculation techniques for next-to-leading order (NLO) have surfaced making possible the generation of processes with many final state particles (up to 6). If NLO calculations are in many cases under control, although not yet fully automatic, even higher precision calculations involving processes at 2- loops or more remain a big challenge.

After a short introduction to particle physics and to the related theoretical framework, we will review some of the computing techniques that have been developed to make these calculations automatic. The main available packages and some of the most important applications for simulation and data analysis, in particular at LHC will also be summarized (see CCP2012 slides [1]).

PARTICLE PHYSICS GOALS AND MEANS

Particle physics targets the study of the ultimate basic elements of matter and of the fundamental forces generated by or acting on them; the smaller the element is, the smaller the probe wavelength must be and, therefore, the larger the interaction energy. Actually, particle physics or more precisely its high-energy physics branch tends to reproduce on earth the range of energies that was prevalent when matter did start to form, at a very early time of the Universe history (figure 1), namely some 10-10 s after the Big Bang. There is a profound duality between the today state of matter probed by highenergy colliders and the state of the Universe shortly after its creation, 13.7 billion years ago.

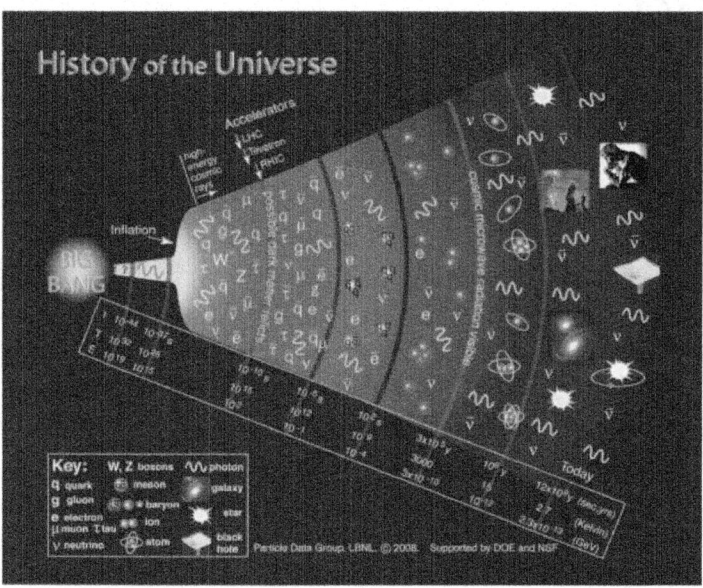

Figure 1: History of the Universe.

The highest man-made particle colliding energy reaches, today, 8 TeV (Tera Electron Volt ~ thousand times the proton mass) on the way to the nominal 14 TeV expected at the large hadron collider (LHC [2]) at CERN [3], Geneva (Switzerland). This energy, however, remains many orders of magnitude smaller than the typical interaction energy occurring in the Universe when quarks, gluons and electrons were moving in a confinement free "hot soup". From this even earlier Universe energy we only have a glimpse, today, through the very high-energy cosmic rays (more than 108 TeV) hitting the upper earth atmosphere. Producing huge particle showers they are detected by large area detectors on earth surface like at the 3000 km2 Pierre Auger Observatory [4] or observed by satellite surveying the earth as proposed by the GEM-Euso Collaboration [5].

Located at CERN, the LHC (figure 2) is the latest built proton-proton collider. It is housed in a 27 km circular, 100 m deep underground tunnel in the Geneva region. The LHC benefits from a large accelerator infrastructure progressively installed at CERN since it was founded back in the 50's. The proton bunches produced from a hydrogen source are accelerated by the Linac and Booster. Then, the PS and SPS increase the energy up to 450 GeV before entering the LHC. Four experiments are located in the LHC ring: two general purpose ATLAS [6] and CMS [7] and two dedicated LHCb [8] for the meson B physics and ALICE [9] for hadronic physics when protons are replaced by heavy ions (e.g. lead-lead collision) in the LHC.

The particle physics overall picture is now embedded in a very solid theory that, shyly enough, we keep calling a Model. The Standard Model (SM) is maybe one of the most developed and rugged theory of any physics fields left alone other research domains. This is probably due to the fact that dealing with fundamental processes, generally, does not lead to the complexity of n-body problems generally intractable, although supercomputers are opening new hopes, at least for small n.

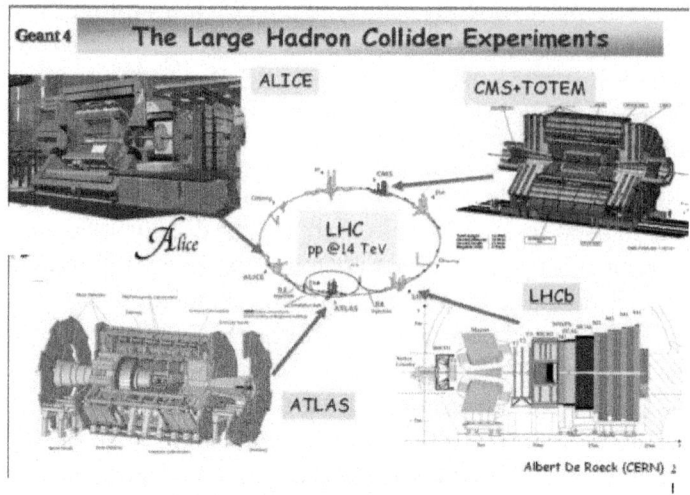

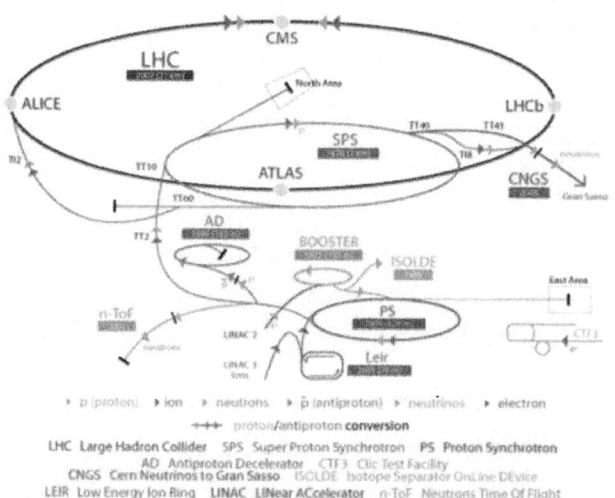

Figure 2: LHC and experiments at CERN.

But that does not mean that the theory is simple when it comes to numerical evaluations. Fortunately methods have been developed based on the underlying quantum field theory to make precise computations. Until now, the experimental data have always either confirmed the theoretical predictions or brought new insight reinforcing the general view.

This success story seems to continue with the recent discovery of a "predicted" boson at the CERN LHC, last July (ATLAS [10], CMS [11]).

Nobody yet dares calling it "Higgs", the last missing ingredient of the SM at current scales. We will soon learn from the experiments whether or not this new boson is the actual SM Higgs proposed as a consequence of a mechanism giving mass to elementary particles [12] [13] [14] [15].

This discovery is materialized by the small bump popping out of the recombined γγ mass histograms (figure 3) for both ATLAS [10] and CMS [11]. These simple plots summarize the extensive computational undertaking behind high-energy physics data analysis. For example, in the CMS experiment, the dots and the error bars represent the observed number of events in a given mass bin. The numbers come from a long chain of data analysis beginning with the trigger event selection, followed by the event reconstruction, the event filtering and analysis for the millions of events recorded by the experiments.

The yellow/green bands represent the best estimation of the so-called background events. They are based on a fit of the data points away from the bump. This fit is also compared to simulated events belonging to background processes. In red, the sum of background and simulated Higgs signal is shown. It is a clear demonstration of the interplay between experimental and simulated data.

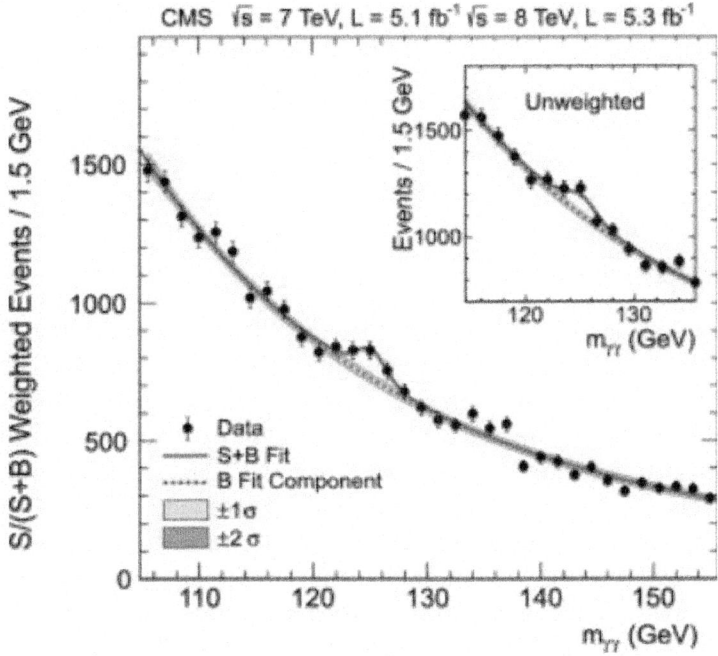

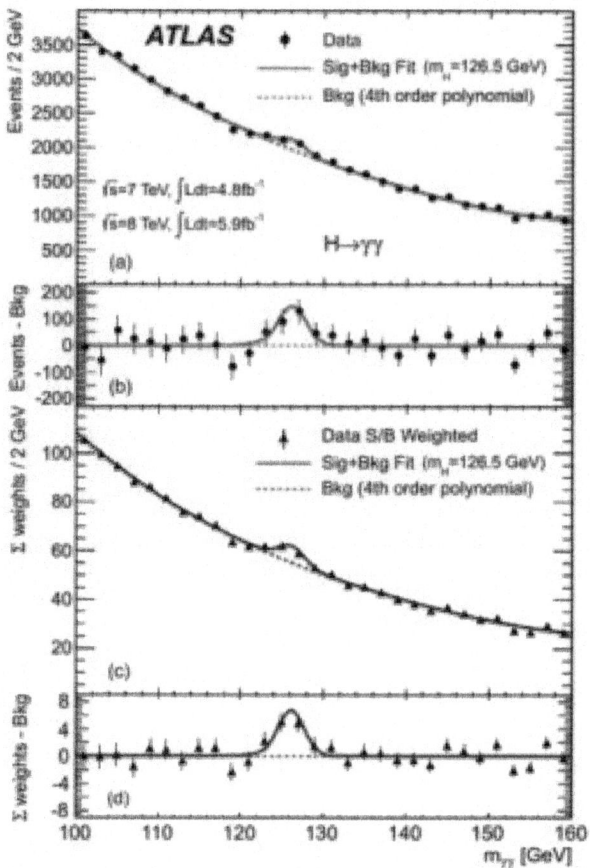

Figure 3: Higgs-like boson signal from CMS and ATLAS (CERN).

Most of the data simulation and analysis is performed on a distributed data grid system called the WLCG (Worldwide LHC Computing Grid) [16]. This infrastructure based on tiers computing centres has been installed all around the world to absorb the heavy computing load of the high LHC statistics. The tier 0 is located at the data production centre (CERN) and absorbs 20% of the load, 11 tiers 1 connected through direct high bandwidth fibre links (10 Gb/s) are scattered across the main participating countries/regions and ~ 140 tiers 2 are providing the specific computing support necessary to research teams. The WLCG provides more than 250,000 processors, and close to 150 PB of disk storage from over 150 sites in 34 countries, producing a massive distributed computing infrastructure that support the needs of more than 8,000 physicists. In section 2, we will describe the general theoretical framework. Section 3 is a short presentation of the perturbative calculus and of the recent higher precision calculation breakthroughs. Then, in the following sections, like Russian dolls, we will go deeper in the experimental simulation chain (section 4): with the description the event formation phases (section 5), of the various steps leading to the building of event generator (section 6), including the automatic calculation of matrix elements (section 7) and their multi-dimensional integration giving the cross-sections (section 8). Section 9 describes the more recent usage of the same matrix element expressions for a more complete event analysis. Finally, section 10 lists some of the computational techniques that are used and shared with other research fields.

PERTURBATIVE AND NOT PERTURBATIVE THAT IS THE QUESTION

Let us first replace this computational activity in the general framework of particle physics and its Standard Model. There are basically two domains of calculation depending on the actual value of the QCD1 coupling constant αs. This constant is actually "running" from small values close to zero in the "asymptotic freedom" regime, namely at high energy when probing the deep inside of hadrons, to large values (αs ~ 1) in the "confinement" region where the partons (quarks and gluons) are kept inside the hadrons.

The Schrödinger equation which represents the time evolution of the quantum state can be approximated by a series expansion in power of αs. This is the so-called perturbative domain. Each term of the series corresponds to a level of precision, the higher the number of terms the more precise the calculation. We will see later also that each term of the series can be graphically represented by a set of "Feynman" diagrams [17], all having similar characteristics. This domain describes the dynamics of QCD interactions. This is the branch of computational particle physics that will be discussed in this paper.

Conversely, if αs is large ~ 1, the hadron quantum system is probed at the confinement limit. Quarks and gluons are all strongly coupled and cannot escape the hadron boundary. Here the interest is more focused on the hadron as a whole. Global values like hadron masses are computed. The perturbative approximation does not apply though and the Schrödinger equation must be solved exactly. However, this has proved to be analytically unfeasible despite many years of trials until K.G. Wilson [18] proposed to replace the continuum phase space by a discrete approximation, the size of the lattice introducing a cut-off stabilizing the solutions. It is only these recent years that this approach became successful with the tremendous increase in computational power of the super-computers and new breakthrough in the basic algorithms. This technique is known as the QCD calculus on lattice or lattice QCD. Please refer to the excellent status report by Karl Jansen in these proceedings.

FROM TREE LEVEL TO HIGHER ORDERS: THE NLO REVOLUTION

Perturbative field theory allows level-by-level calculation of particle interactions. The technique was developed by R.P. Feynman [19] in the 50's. Each precision level can be matched to a set of similar diagrams. Let's take a simple example, in QED2 (figure 4), at the lowest level or tree level in the coupling constant α, the simple $e^+e^- \rightarrow e^+e^-$ process involves only 2 diagrams: a s-channel graph when the electron and the positron annihilate to form a virtual photon, also called propagator, which then decay into an electron and a positron and a t-channel where a photon is exchanged between the electron and the positron. In QED, the coupling constant is small enough ($\sim 1/137$, although also running) so that in most cases first order calculations provide good enough precisions matching experimental accuracies. The input and output particles are called "legs".

But in some cases, experimental errors are much smaller than tree level computations like at the precision e^+e^- colliders. They can even be orders of magnitude smaller when, for example measuring the muon anomalous magnetic moment factor (g − 2)/2 at 0.14 ppm precision [20]. The theoretical precision should be, at least, as good and computing the next terms in the expansion series becomes necessary both for electroweak and hadronic corrections.

The next level in precision is the so-called "Next-to-Leading Order" (NLO). It essentially implies adding 1-loop corrections. A photon can be exchanged between the initial state electron and positron, an e^+e^- loop can be inserted in the propagator or even (not shown); a second photon is exchanged in the t-channel (creating box diagrams). But the virtual exchanged particle can also be real and diagrams with an external photon must be included. If one sticks precisely to the same final state, the additional photon should not be observed namely, it should be either soft or almost collinear with the emitted electron. Loops and/or legs can be put on all elementary parts of the tree-level diagrams and the number of diagrams increases dramatically. Next-to-Next-Orders calculations follow the same trend, adding more and more diagrams and therefore becoming more and more computational intensive.

At LHC, QCD plays a major role and things get more complicated. First the proton is not an elementary particle, but a complex system of 3 valence quarks in a gluon fog and a sea of quarkantiquark pairs. At LHC energy, the interest is on the hard scattering of proton constituents, namely gluon-gluon, quark-gluon or quark-quark. Therefore many different initial states must be taken into account. The selection of the initial partons within the proton is made through the use of the so-called parton distribution functions (PDF) which describe the parton content of the proton. These functions are mainly parameterizations based on data from the e-p experiments at HERA (DESY [21]) or in $p\bar{p}$ experiments at D0/CDF (Tevatron [22], Fermilab) and now at LHC.

Second, the final states may involve many particles, i.e. many legs, like shown in the experimental wish list discussed at the "Les Houches" meetings [23]. This is a huge step in complexity compared to the 2→4 channels studied at the electron-positron collider (LEP II [24]).

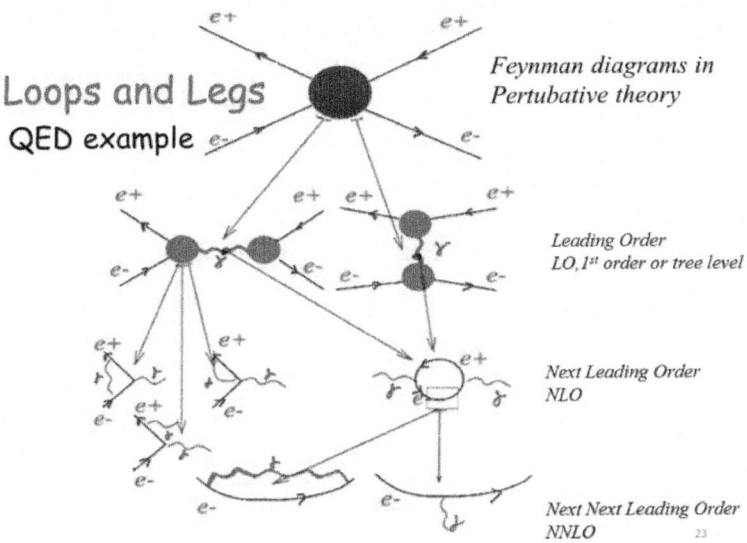

Figure 4: Loops and legs in QED.

Finally, contrary to QED where tree-level calculations are often a good approximation, QCD estimations must go beyond LO which depend too much on arbitrary scales. The renormalization scale μ_R enters in the cross-section as well as in the value of α_s. Renormalization is a necessary operation to cancel infinities arising in the calculation of the cross-section due to UV (Ultra-Violet) divergences. The factorization scale μ_F which also enters in the cross-section as well as in the parton distribution function is related to the soft and collinear divergences, namely in the IR (Infra-Red) region. These scales are arbitrary and induce large normalization errors (10-20% or even more). Higher order calculations reduce largely the sensitivity to these arbitrary scales and this trend is even more pronounced when the number of legs is large.

QCD NLO calculation of many legs processes may lead to more than 105-6 diagrams and for each, the amplitude or matrix element can be so large in the usual Feynman representation that computer memory sizes become an issue. Clearly the Feynman diagrammatic approach was not quite suited for this kind of calculation.

Although this conventional method is still in use and even being improved, new approaches originated from the string/twistor theory have recently been developed for the calculation of one-loop processes. The on-shell unitarity method [25] to the expense of rewriting the quantum field theory in terms of invariant gauge on-shell intermediate expression simply proposes to cut the loop diagrams in one or many places decomposing the initial one-loop calculation to a set of tree level amplitudes, easily handled by the conventional tree-level diagram calculators. Following these ideas many developments have been undertaken (see for a summary until 2008 [26]) and have led to automatic calculation of 2→4 or 2→5 NLO processes (figure 5).

Even if these techniques for the most complex calculations have not reached full automation, they have made the computation possible with the current computers. Event generators at NLO based on SHERPA [27] and BlackHat [28], MadGraph [29] and others have been heavily used by the LHC experiments.

NLO Revolution

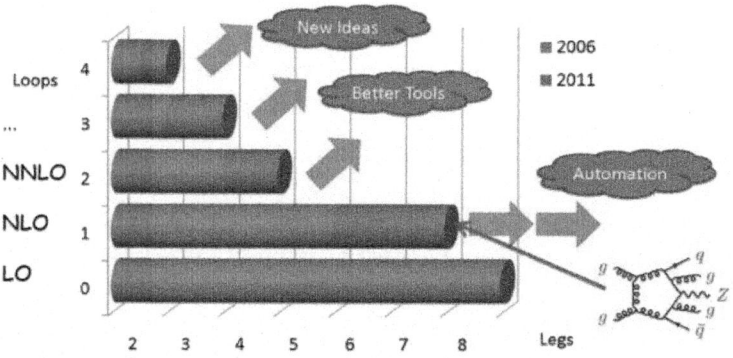

Figure 5: NLO Revolution [30].

EVENT SIMULATION AND DATA ANALYSIS

It is impossible to calculate analytically the detector responses to an event produced in the quite complex LHC experiments. One therefore relies on

Monte-Carlo simulations. Any process of interest can be actually generated and simulated in a virtual representation of the detectors. The simulation is supposed to mimic in detail what is detected when beams collide. Figure 6 presents schematically the various steps of the simulation chain. On the top, the "physics model" describes the framework in which the calculation is performed. The basic model is the Standard Model (SM). But extension or beyond SM models are also implemented like for example the supersymmetric (SUSY) model. This model suggests that any known fermion (boson) has a partner boson (fermion) yet unobserved due to its high mass. Clearly beyond standard model (BSM) theories are made up to cure problems arising in the SM beyond current energy, in particular, the divergent contributions to the Higgs mass.

The actual process is, then, selected by defining the input and output particles as well as the order of calculation. The matrix element or the expression of the process probability is then automatically prepared based on the physics model data. After integrating of the matrix element over the parameters phase space, the code for an event generator is constructed. It will generate random samples of energy-momentum four-vectors for all final state physical particles.

Each generated particle is then propagated into a model representing the experiment built by a detector simulation package. GEANT [31], initiated at CERN and developed by an international collaboration is the main package used by our community. Version 4 is an extremely detailed description of the detector components down to nuts and bolts and read-out cables. All physics particle-matter interactions have been implemented. Each particle, step by step is tracked in the material forming the structures as well as the detection sensitive elements.

The charged particles may produce primary gas ionization, scintillation or Cherenkov light depending on the materials. Photons/gammas create electromagnetic showers producing hundred to thousand low energy particles which are all tracked down to final absorption. This information is collected by sensitive elements: wires, silicon strips/pixels or photo-detectors. All particles are followed until stopping in material or escaping the experimental setup. The huge amount of information produced by the detector simulation should in principle represent exactly what a real event in a real detector would produce. Obviously, confidence on the detector simulation accuracy has been obtained after thousands of validation measurements for each type of particles, materials and detectors.

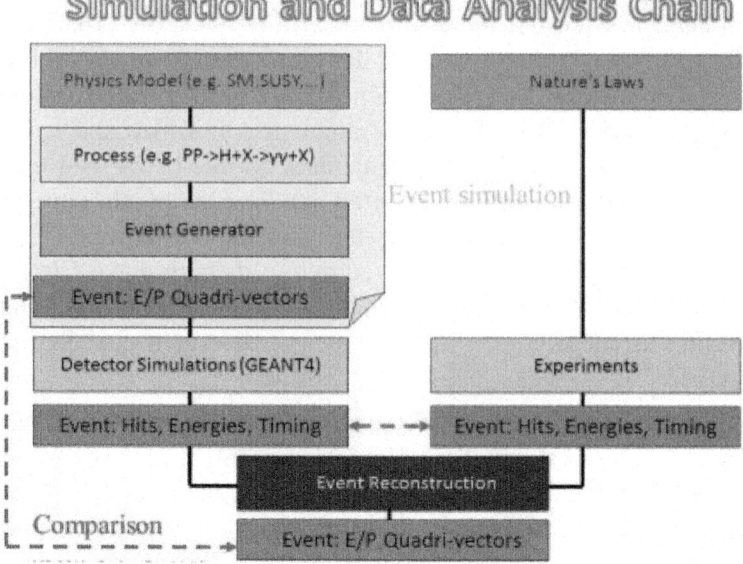

Figure 6: Monte-Carlo simulation and the data analysis chain.

At this level, one has now simulated i.e. theory "generated" event data that can be compared to the experimental data produced by Nature in the physical detector. Both simulation and real data are then reconstructed by the same reconstruction programme to get the initial set of energy-momentum fourvectors representing the actual simulated or observed event. Comparing the reconstructed simulated four-momenta to those produced by the event generator is a test of self-consistency of the whole sys-tem although departure between the simulated and the experimental data may give hints to new physics discovery.

The final analysis is generally done with the Root data analysis toolkit [32], developed at CERN. Events data are put in large ntuples (sort of expandable spread sheets) on which filters and fits can be applied. Estimators can then be computed and placed in histograms. This is how plots like those shown in figure 3 have been produced. Would the experimental data have been precisely fitted by the green-yellow background band, we would had concluded to a no Higgs discovery and higher limits on its production would had been set. But hopefully some excess was found, signalling the existence of something beyond the expected background, a new particle, maybe the expected Higgs.

Event Sequential Formation Stages

Let's now focus on the physics of an event formation. In a high-energy *pp* scattering, the generation of a single event (figure 7) takes several steps. First, 2 partons (flavour and colour) are selected in each of the colliding protons. Before interacting, partons are allowed to emit gluons (initial state radiations). Then the actual hard-scattering interaction occurs and 2 or more particles are produced. These particles may then decay like in heavy quark/leptons or bosons (Z, W) producing several daughters. Finally, the last remaining quark and gluon undergo partonic showers before the final hadronization step occurs, namely the recombination of end partons in hadrons (baryons or mesons).

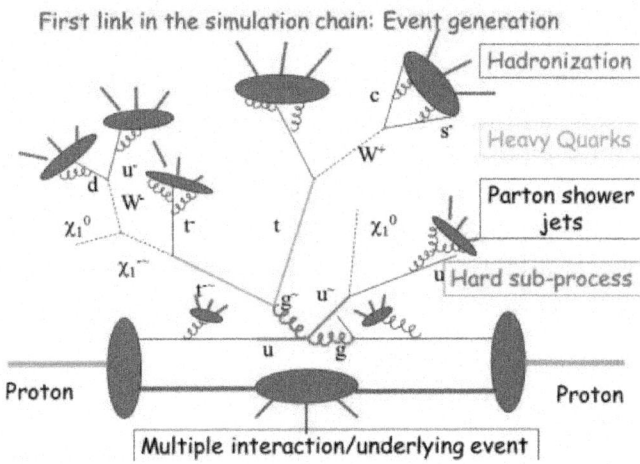

Figure 7: Sequential steps in the formation of an event. The event presented here is a SUSY event with a hard process: u-quark gluon →u-squark gluino, the u-squark being the superpartner of a uquark and the gluino the gluon counterpart.

In addition, several interactions between different partons can occur in the same proton collision leading to multiple parton interactions (MPI). Finally, in the collision of proton bunches more than one collision can occur and multiple vertices are created, this is called the underlying events (UE). The odds that several interactions producing a hard scattering each leaving, in the detectors, high momentum transfer particles is quite small, but these extra collisions produce many additional tracks and energy depositions in the detector. These parasitic effects make the analysis more difficult as each track or energy deposition should be associated to the right vertex and, therefore, the whole reconstruction process takes more CPU resources and becomes more error prone.

PROCESS CALCULATION

Now that the physics of an event formation has been described, let us see how this is implemented in the global process calculation (figure 8). A flurry of packages has been developed to carry out the full event simulation.

Based on the model Lagrangian or, more generally, on the group symmetries and fields definitions, symbolic manipulation languages like Reduce, Maple, Mathematica or Form [33] are used to encode in a model file the various couplings and the propagators expressions either on an analytic form or as a set of numerical libraries. The Form language has been developed to handle very large expressions buffered on disk files. Specific programs like LanHEP [34], FeynRules [35] or SARAH [36,37] (SUSY) have made this derivation automatic.

Then a specific process can be selected. The initial particles undergoing the hard collision are generated after initial state radiation and beamstrahlung (CIRCE [38]) for electrons or parton extraction from the parton distribution functions (PDF) (LHApdf [39], CTEQ [40], MRST [41],) for the protons has been applied. One of the automatic amplitude/matrix element generators can be put in action to produce the global expression of the process integrand.

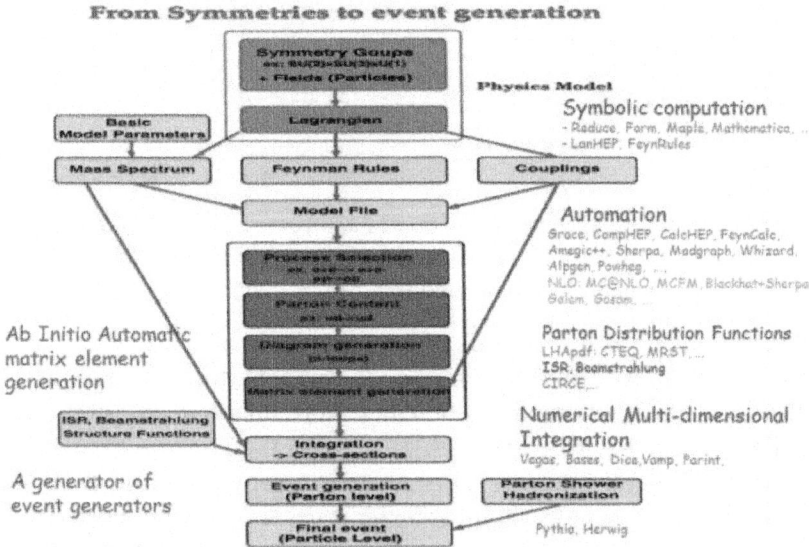

Figure 8: From symmetries to Event Generation.

This expression must then be integrated over the full phase space, namely over all possible legs momenta/energy, only constrained by the energy momentum conservation and the initial conditions. If loops are involved, integration over the loop propagators must also be performed. The multidimensional integration provides the total cross-section and the probability function for the generation of the events.

The event generator can, now, be constructed. It will randomly generate events at the parton level. Parton shower and hadronization programs like Pythia [42] or Herwig [43] will take over to produce the final full-fledged event ready to enter the detector simulation stage.

MATRIX ELEMENT AUTOMATIC CALCULATIONS

The matrix element calculation it-self has been for many years performed manually by talented theorists. Back in the 80's, a group in Japan from the KEK laboratory (Grace) [44] and a Russian group from Moscow State University (CompHEP) [45] [46] started to develop packages to do these lengthy and error prone calculations automatically, that was the beginning of the automatic process calculation endeavour.

The automation was initially seen as a toy project, at best to train students. But the experimental requests for new process calculations went exploding as well as the complexity of the process calculation and, finally, the automatic approach became the baseline. The first event generator with an automated computed matrix element was probably $\text{grc4f } e^+e^- \to 4$ fermions [47] which was used for the LEP2 analysis. For the first time, spin correlations and mass effects were included. Nowadays most calculations are made by computer tools in a combination of symbolic and numerical operations. The various steps of the automatic calculation flowchart, at tree level can be summarized as follows:

- Enumeration and description of the set of diagrams entering the calculation of a given process matrix element (e.g. standalone QGRAF [48] or included in the automatic packages e.g. [44], [45])

- Preparation of the matrix element expression for each diagram by various calls to the vertex and propagator subroutine library or by building a global analytical expression.

- Multi-dimensional integration over the full phase space (including the PDF variables) to produce 1) the total cross-section 2) the probability function describing how likely an event with a given set of input and output four-vector will occur.

- Preparation of the event generator code. According to the probability function, random set of the process four vectors will be generated.

- Inclusion of the final state radiation calls as well as the parton shower-ing and hadronization steps based on Pythia [42] or Herwig [43].

For reference here follows some tree level matrix element generators: GRACE [44], CompHEP [45], CalcHEP [49], FeynCalc [50], MadGraph [29], AMEGIC++ [51] as well as multi-channel event generators HELAC-PHEGAS [52], ALPGEN [53], GR@PPA [54], O'Mega [55], Sherpa [27], Whizard [56], Pythia [42], Herwig [43]. In practice there are, yet, no complete and fully automatic matrix element generators at NLO, but many packages are being developed in this direction. Recent reports can be found in [57,58,59] and the main contenders are: MCFM [60], NLOJET++ [61], BlackHat [62], Rocket [63], CutTools [64], MadLoop [65], OpenLoops [66], GOLEM [67], POWHEG [68], aMC@NLO [69,70], Sherpa+BlackHat [71], MadGOLEM [72], GoSam [73], HELAC-NLO [74].

MATRIX ELEMENT MULTI-DIMENSIONAL MONTE-CAR-LO INTEGRATION

Unfortunately matrix elements tend to have singularities in some parts of the phase space making their integration difficult: for example, in the collinear region where two particles (e.g. an electron and a photon) get a small angular separation. This corresponds to the quantum physical situation where an electron alone cannot be distinguished from an electron closely accompanied by a soft photon. In that case one gets a singularity that can be tamed by introducing a cut on the photon-electron angle smaller than the actual experimental angular resolution.

Anyway singularities occur in the expression of the integrand and the skill of the event generator developers are still needed to craft the most efficient variable transformations so that singularities are avoided or minimized. The mapping between integration and physical variables is coded into the "kin-ematics subroutines". They are selectively plugged into the integrand code according to its sensitivity to specific divergences.

For the less acute variations, different techniques automatically adapting the sampling grid to the shape of the integrand have been implemented in the main multi-dimensional integration packages like Vegas [75,76], Bases/Spring [77], Dice [78], Foam [79], Vamp [80], Mint [81], ParInt [82]:

- Importance sampling: the grid intervals or bins get smaller and therefore the sampling is denser in regions of high integrand values as this phase space regions will contribute most to the total integral.
- Stratified sampling: the bins get smaller in the region of fast variation

of the integrand so that from one interval to the next the function varies slowly and smoothly.

• Multi-channel sampling: several mappings, namely grid definitions, can be selected depending on the phase space region. The method is to always select the mapping that suits best the behavior of the integrand at the particular sampling point.

For processes up to 2→4, at tree level the full automation is now usual business. For more particles in the final state (e.g. 2→6), the integration stage may take a lot of CPU time and the calculation may become intractable if the kinematics mappings are not adequate. For NLO calculations the number of integration dimensions increases due to the additional loop four-momenta and the integration difficulties become even more stringent.

In addition, the size of the integrant can be enormous and compiler and computer memory limitations become an issue. Fortunately after the NLO revolution, some computations have become feasible, but automation has not yet been fully implemented.

When the integrand becomes too large, it can be cut in smaller parts (e.g. each sub-diagram) distributed over many nodes. This is one direction for parallelization. A second one to accelerate the computation is to share the whole phase space again over many nodes. So that, in the end, each node will compute a part of the integrand for a part of the phase space. However, the integration algorithm requires the full value of one or several integrand calculations over the whole phase space, before executing the next iteration. This is the serial part of the integration algorithm and load balancing between the nodes becomes an issue that has to be addressed [83,84].

Recently, it has been shown that the computation of loop integrals or tree level processes on GPUs have led to dramatic acceleration [85,86]. Supercomputers will certainly benefit these calculations.

MATRIX ELEMENT METHOD FOR DATA ANALYSIS

As we have seen, the process matrix element is at the heart of the event generator and its calculation can be made automatic in most cases, at least, at tree level. But a new application of the matrix element is looming up for the data analysis it-self.

The usual way of extracting new insights from the experiments is by comparing the simulated and experimental event distributions. This is the conventional approach. A more elaborate, but similar technique called the template analysis has been developed along these lines and involving likelihood maximization. But a new approach is nowadays preferred. It involves the use of the matrix element expression in the event selection algorithm it-self. This is the matrix-element method [87] or also called the matrix element likelihood algorithm (MELA) in CMS.

Template method

Let us assume that a process parameter is to be estimated from experimental data. Let's take for example, the mass of the top quark. The t-quark has a very short lifetime and decays, even before it gets hadronized, in various channels including in lepton + jet. From these decay products the top quark mass can be reconstructed. The template method requires the simulation of several similar distributions assuming different top masses. Then, a likelihood expression is built and its maximization provides the best fit between the experimental and simulated distributions. This method is straight forward, but only a small part of the full event information is used, namely the reconstructed top mass.

The rest of the event data like the distributions of remaining particles and their correlations does not enter this optimization.

Matrix-element Method

From the same selected sample, a probability is built for each event taking into account all final state particle kinematics to estimate to which category it is most likely part of. The following probability P_{sig} is therefore estimated event per event:

$$P_{sig}(x, m_t) = \frac{1}{N} \int_{y, q_1, q_2} d\Phi(y) \, |M_{t\bar{t}}(y, m_t)|^2 \, W(x, y) \, f_{pdf}(q_1) \, f_{pdf}(q_2) \, d\,q_1 dq_2$$

where:

- x represents the measured kinematic variables for all final state particles

- m_t is the parameter to be estimated, here the top mass

- $|M_{t\bar{t}}|^2$ is the square of the matrix element associated to the signal process under study, here the production of a pair of top quarks.

- $W(x,y)$ is the transfer function mapping the measured four-vectors and the matrix element theoretical parameters.

- $f_{pdf}(q)$ is the parton distribution function described above giving the probability that the initial colliding partons with q_1 and q_2 four-momenta are produced by the interacting protons.

This expression is integrated over the phase space and normalized to the observed cross-section N. Similar probabilities $P_{bkg1...n}$ are formed assuming that the same event belongs to background processes, namely processes having a similar final state but not through the decay of top quarks.

If the $f_{sig,bkg1...n}$ are the fractional parts of the total sample (signal or backgrounds), a global probability for one event with the kinematics x to form a mass m_t whether from signal or from background can then be computed:

$$P_{evt}(x, m_t) = f_{sig} \, P_{sig} + f_{bkg1} \, P_{bkg1} + \cdots + f_{bkgn} \, P_{bkgn} \, .$$

Finally, a sample level likelihood is computed as the product of the global probability of all events and the following inverse log is minimized to get the most probable m_t:

$$\mathcal{L}(m_t) = -ln(\textstyle\prod_{evt} P_{evt}(x, m_t)) \, .$$

In this method implemented in the Mad Weight [88] package, the whole event information is incorporated in the parameter estimation. In addition the separation between the experimental data and the theory input is well identified. But this is a very CPU time consuming procedure as for each event several multi-dimensional integrations are necessary. To help solving this problem, the use of GPU has been implemented successfully [89].

The template (figure 9) and the matrix element method (figure 10) have been used with great success for establishing the value of the top mass at the Tevatron by the D0 [90] and CDF [90] experiments.

A similar technique has also been used at LHC to estimate the Higgs mass. A global expression MELA is computed from the likelihood, namely MELA is 0 (1) if the probability of the event to be a background (signal) is maximal. Each event is presented on a scatter plot with the estimated Higgs mass on the x-axis and the MELA value on y-axis. Setting the threshold on MELA above 0.5 selects events more likely to be a signal than a background which greatly enhances the signal over noise ratio of the Higgs mass measurement (figure 11).

Matrix element methods incorporating NLO corrections have also been studied [91], but have not yet been implemented for experimental data analyses.

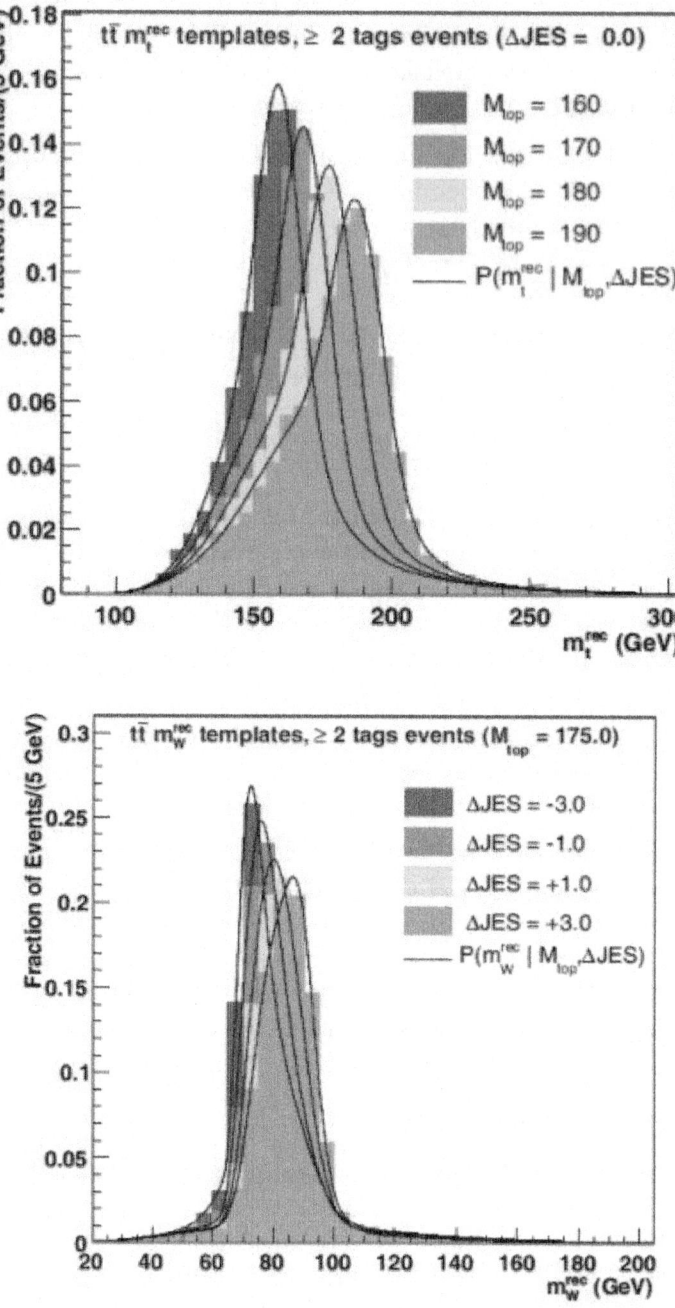

Figure 9: The template distributions for reconstructed top mass (left) and mw (right) for CDF [90]

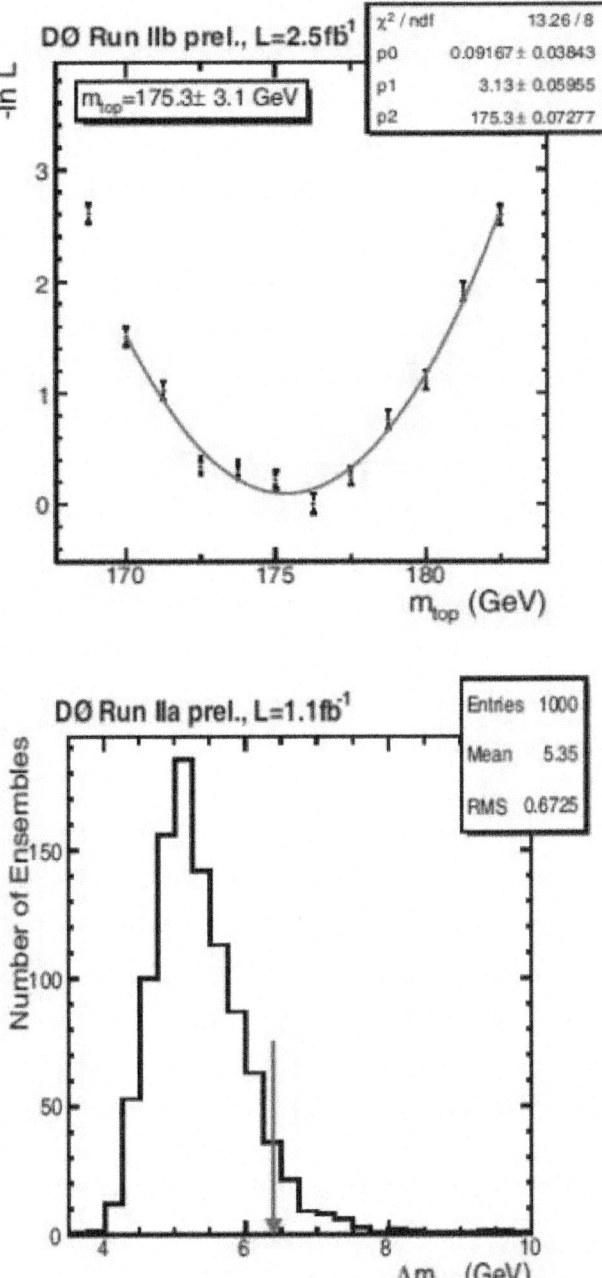

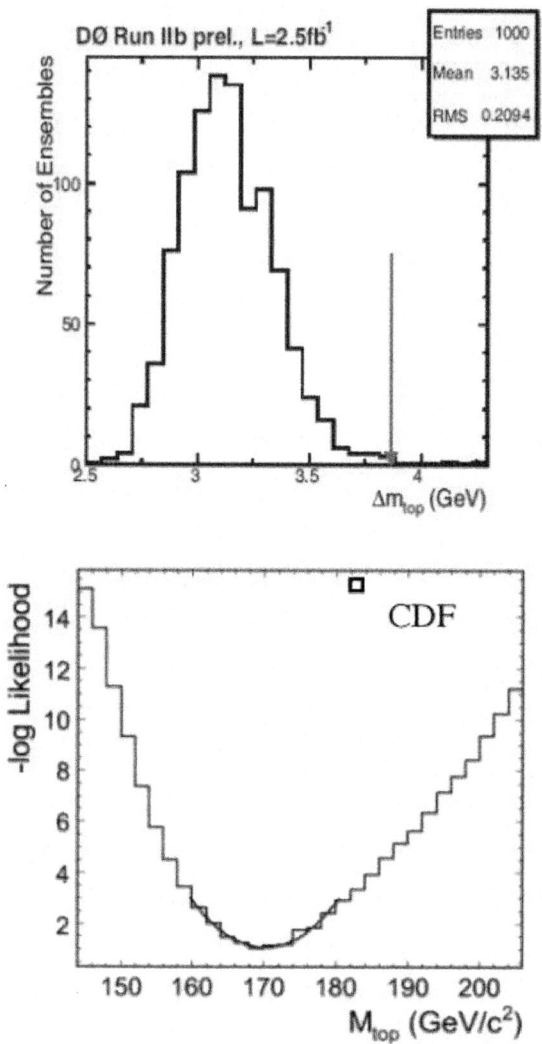

Figure 10: Measured likelihood for the top mass at D0 and CDF [90,92].

Cross-Disciplinary Fertilization and Conclusions

Computational particle physics covers two main branches, one dealing with the non-perturbative calculations, lattice QCD, which is virtually intractable without the use of large parallel super computers. Recent developments show that thanks to a new algorithm and a drastic increase in computing power, the hadronic mass spectrum and other global parameters are now well reproduced.

The other branch is related to the perturbative region probing the inner part of the nucleon as well as its high-energy dynamics. These computations, for the highest available energy, have also become feasible thanks to new NLO calculation methods and, again, to the availability of a large Grid computing infrastructure. The automation is now on track for the most complex calculations. Experimental design, physics analysis and theoretical interpretation at current colliders deeply rely on the development of these "new" tools providing matrix elements and event generators.

In fact, perturbative calculations both benefit from and motivate the development of techniques and methods common to other domains in computational science including:

- Symbolic calculation: the traduction of symmetries and fields to Feynman rules and couplings has been automated using algebraic languages like Reduce, Maple or Mathematica. Often created to support these calculations, they are now multi-purpose languages and are heavily used in many other research fields. However, the handling of very large expression of high complexity has triggered the implementation of expression disk caching and parallel techniques like in the language Form.

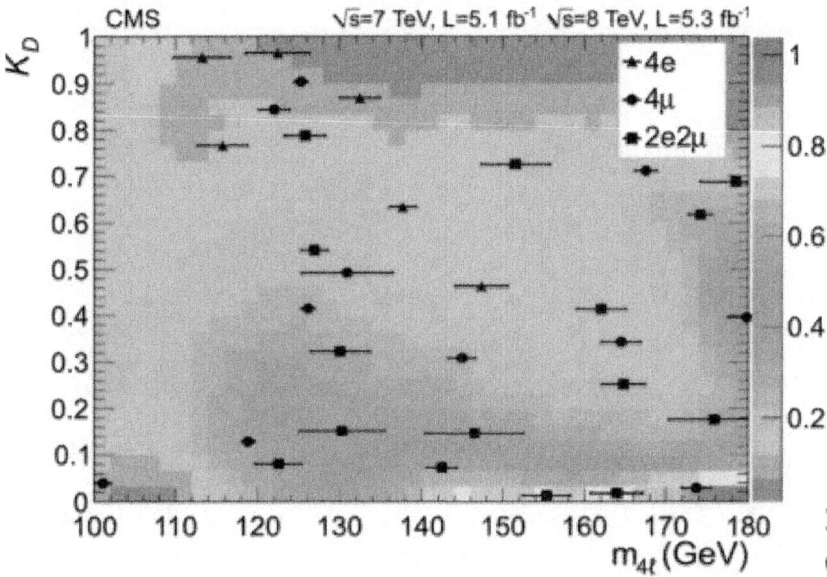

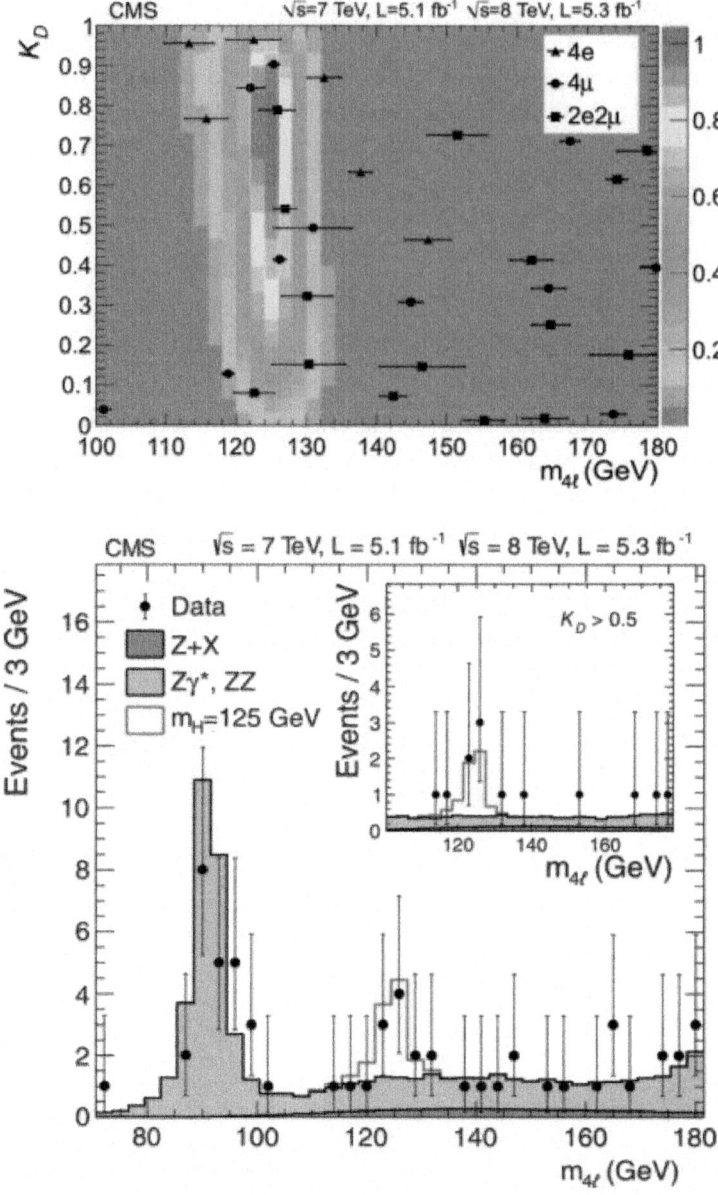

Figure 11: CMS: MELA analysis of Higgs decaying in 4 leptons (from [11]).

- Parallelism:

  o Numerical multi-dimensional integration: one of the main issues in perturbative computation is the integration over many dimensions of very large singular expressions. Some of the integration packages have been made parallel and able to run on GPU or supercomputers.

o Process calculation factorial growth: the number of models, of processes and sub-processes is sky rocketing prompting for the use of massively parallel supercomputers or large clusters.

- High precision floating point computation: often double precision is not sufficient. Quadruple or octuple precision becomes necessary to achieve convergence for the most singular integrands leading to order of magnitude increase in computing time [93]. Specific libraries and hardware developments based on co-processors or GPUs [94] are in progress.

- Large computing infrastructures:

o The internationally distributed Data Grid architecture, mostly created to cover the collider experiments needs, has fulfilled the expectations. It gave birth to the "Cloud" technology, a new commercial endeavor.

o Although most of the current data analysis and simulations programs are serial, both the manycore/GPU long term industrial trends and the need for higher precision theoretical predictions are incentives to implement multi-threading and parallel execution techniques (e.g. for analysis [95] and simulation [96]). Certainly not an easy task in object oriented environments, but it will open access to the huge power of supercomputers.

- General data analysis toolkits (e.g. Root) implementing advanced statistical methods developed for particle physics has found applications in other environment like biology, finance and medicine.

Finally, the initial dream of automatic perturbative physics calculations has triggered the development of a flurry of packages and specific tools. Some are at the heart of today high-energy colliders and astroparticle experiment data analysis. But these packages have usually very little in common. Despite great attempts to standardize some aspects like the output files format (StdHep [97], HepMC [98]) or the interface with NLO calculations (BLHA [99]), they still have no common input syntax, no general parameter definition format, a poor modularity, no universal submission scheme, no common visualization and analysis tools, and so on. It is time to address these issues to facilitate their necessary detailed comparisons and their embedding in the large experimental simulation packages.

REFERENCES

1.   Perret-Gallix D 2012 CCP 2012: Computational particle physics for event generators and data analysis (slides) http://acpp-coll.in2p3.fr/cgibin/twiki.source/pub/ACPP/PresentationsNotes/ccp_2012-3.pptx

2.   LHC: Large Hadron Collider http://public.web.cern.ch/public/en/LHC/LHC-en.html

3.   CERN: The European Center for Nuclear Research http://www.cern.ch

4.   Pierre Auger Observatory http://www.auger.org/

5.   GEM-EUSO Proposal http://jemeuso.riken.jp/en/index.html

6.   ATLAS experiment at CERN LHC http://atlas.ch/

7.   CMS experiment at CERN LHC http://cms.web.cern.ch/

8.   LHC-B experiment at CERN LHC http://lhcb.web.cern.ch/lhcb/

9.   ALICE experiment at CERN LHC http://aliceinfo.cern.ch/

10.   ATLAS Collaboration 2012 Observation of a new particle in the search for the Standard Model Higgs boson with the ATLAS detector at the LHC Phys. Lett. B, 716 p 1-29

11.   CMS Collaboration 2012 Observation of a new boson at a mass of 125 GeV with the CMS experiment at the LHC Phys. Lett. B 716 no.1 p 30-61

12.   CERN web site. The search for the Higgs boson http://public.web.cern.ch/public/en/science/higgs-en.html

13.   Brout R and Englert F, Broken Symmetry and the Mass of Gauge Vector Mesons 1964 Phys. Rev. Lett. 13 no. 9 p 321-323

14.   Higgs P 1964 Broken Symmetries and the Masses of Gauge Bosons Phys. Rev. Lett. 13 508–509

15.   Guralnik G S, Hagen C R and Kibble T W B 1964 Global Conservation Laws and Massless Particles Phys. Rev. Lett. 13 585–587

16.   WLCG: Worldwide LHC Computing GRID http://wlcg.web.cern.ch/

17.   Kaiser D, Physics and Feynman's Diagrams 2005 American Scientist 93 156 http://web.mit.edu/dikaiser/www/FdsAmSci.pdf

18.   Wilson K G 1974 Confinement of quarks Phys. Rev. D 10 2445

19.   Feynman R P 1949 Space-time approach to quantum electrodunamics Physics Review 76 769- 789

20.   E-898, Fermilab, The New (g – 2) Experiment: A proposal to measure the Muon Anomalous Magnetic moment to +- 14 ppm precision http://gm2.fnal.gov/public_docs/proposals/ProposalAPR5-Final.pdf

21. HERA at DESY (Hamburg) http://www.desy.de/research/facilities/ hera_experiments/index_eng.html

22. TEVATRON at FERMILAB http://www.fnal.gov/pub/science/ experiments/energy/tevatron/

23. Les Houches Working group, The SM and NLO MultiLeg and SM MC Working Groups 2010 arXiv:1203.6803.

24. LEP at CERN (Geneva) http://public.web.cern.ch/public/en/research/ lep-en.html

25. Bern Z, Dixon L J, Dunbar D C and Kosower D A 1995 Fusing gauge theory tree amplitudes into loop amplitudes Nucl. Phys. B 435 59–101

26. Les Houches Working group, The NLO MultiLeg Working Group: Summary Report 2008 arXiv:0803.0494

27. SHERPA Collaboration, SHERPA http://sherpa.hepforge.org/trac/wiki

28. Bern Z, Diana G, Dixon L, Febres Cordero F, Hoeche S et al., Four-Jet Production at the Large Hadron Collider at Next-to-Leading Order in QCD 2012 Phys. Rev. Lett. 109 042001

29. Alwall J, Herquet M, Maltoni F, Mattelaer O and Stelzer T 2011 MadGraph 5: Going Beyond JHEP 1106 128

30. Glover N and Heinrich G 2011 Computations in Particle Physics: in ACAT' 2011 Uxbridge, London UK (Advanced Computing and Analysis technologies) http://indico.cern.ch/getFile.py/access?contribI d=124&sessionId=15&resId=0&materialId=slides &confId=93877

31. GEANT 4 Collaboration. Geant 4 http://geant4.cern.ch/

32. Root Collaboration. Root http://root.cern.ch/drupal/

33. Kuipers J, Ueda T, Vermaseren J A M and Vollinga J, FORM version 4.0. arXiv:1203.6543

34. Semenov A 2010 LanHEP - a package for automatic generation of Feynman rules from the Lagrangian. Updated version 3.1 arXiv:1005.1909

35. Christensen N D and Duhr C 2009 FeynRules - s - Feynman rules made easy Comp. Phys. Comm. 180 1614-1641

36. Staub F 2012 SARAH 3.2: Dirac Gauginos, UFO output, and more Comp. Phys. Comm. 184 1792

37. Staub F, Ohl T, Porod W and Speckner Ch 2012 A tool box for implementing supersymmetric models Comp. Phys. Comm. 183 2165 http://sarah.hepforge.org/Toolbox.html

38. Ohl T Beam spectra: CIRCE, Luminous, Pandora http://theorie.physik.

uniwuerzburg.de/~ohl/lc/generators-sections013.html#circe

39. Whalley M and Buckley A LHAPDF http://lhapdf.hepforge.org/

40. CTEQ Collaboration. The Coordinated Theoretical-Experimental Project on QCD http://www.phys.psu.edu/~cteq/

41. Martin A, Stirling J, Thorne R and Watt G MRS/MRST/MSTW Parton distributions http://durpdg.dur.ac.uk/hepdata/mrs.html

42. Sjöstrand T et al. Pythia http://home.thep.lu.se/~torbjorn/Pythia.html

43. Webber B et al. HERWIG http://www.hep.phy.cam.ac.uk/theory/webber/Herwig/

44. Kaneko T, Kawabata S and Shimizu Y 1997 Automatic generation of Feynman graphs and amplitudes in QED Comp. Phys. Comm. 43 279-295 http://minami-home.kek.jp/

45. Boos E E, Dubinin M N, Edneral V F, Ilyin V A, Kryukov A P, Pukhov A E, Rodionov A Ya, Savrin V I, Slavnov D A and Taranov A Yu 1989 CompHEP: Computer system for calculations of particle collision characteristics at High Energy Moscou State University: Institute of Nuclear Physics, Preprint 89-63/140 http://comphep.sinp.msu.ru/ http://ccdb5fs.kek.jp/cgibin/img/allpdf?199003342

46. Boos E E and Dubinin M N 2010 Problems of automatic calculation for collider physics Physics Uspekhi 53 1039-1051

47. Fujimoto J, Ishikawa T, Kaneko T, Kato K, Kawabata S, Kurihara K, Munehisa T, Perret-Gallix D, Shimizu Y and Tanaka H 1997 Grc4f v1.1: A Four fermion event generator for e+ ecollisions Comp. Phys.Comm. 100 128-156

48. Nogueira P, Automatic Feynman graph generation 1993 Journal of Computational Physics 105 279-289 http://cfif.ist.utl.pt/~paulo/qgraf.html

49. Puhkov A CalcHE http://theory.sinp.msu.ru/~pukhov/calchep.html

50. Mertig R FeynCalc http://www.feyncalc.org/

51. Krauss F, Kuhn R and Soff G 2002 AMEGIC++ 1.0, A Matrix Element Generator In C++ JHEP 0202 044

52. Kanaki A and Papadopoulos C G, HELAC-PHEGAS: automatic computation of helicity amplitudes and cross sections http://arxiv.org/abs/hep-ph/0012004

53. Mangano M L, Moretti M, Piccinini F, Pittau R and Polosa A 2000 ALPGEN V2.14: A collection of codes for the generation of multi-parton processes in hadronic collisions in Workshop on Computer Particle Physics: (CCP 2001) Automatic Calculation for Future Colliders

arXiv:hep-ph/0012004

54. Odaka S and Kurihara Y 2012 GR@PPA 2.8: initial-state jet matching for weak boson production processes at hadron collisions Comp. Phys. Comm. 183 1014

55. Moretti M, Ohl T and Reuter J 2001 O'Mega: An Optimizing matrix element generator in 2nd ECFA/DESY Study 1998-2001 1981-2009 hep-ph/0102195. http://theorie.physik.uniwuerzburg.de/~ohl/omega/

56. Kilian W, Ohl T and Reuter J 2011 WHIZARD: Simulating Multi-Particle Processes at LHC and ILC Eur. Phys. J., no. C 71 1742 http://whizard.hepforge.org/

57. Binoth T 2008 LHC phenomenology at next-to-leading order QCD: theoretical progress and new results in 12th International Workshop on Advanced Computing and Analysis Techniques in Physics Research (ACAT 2008) arXiv:0903.1876 POS (ACAT08) 011

58. Maitre D 2010 Automation of Multi-leg One-loop virtual Amplitudes ,POS (ACAT2010)008 http://pos.sissa.it/archive/conferences/093/008/ACAT2010_008.pdf

59. Tramontano F 2012 Progress in automated Next-to-Leading-Order calculations in ACAT 2011: Advanced Computing and Analysis Technologies for Research, J. Phys. COnf. Ser. 368 012058

60. Campbell J, Ellis K and Williams C. MCFM - Monte Carlo for FeMtobarn processes http://mcfm.fnal.gov/

61. Nagy Z 2003 Next-to-leading order calculation of three jet observables in hadron hadron collision Phys.Rev. D 68 094002 http://www.desy.de/~znagy/Site/NLOJet++.html

62. Berger C F, Bern Z, Dixon L, Febres Cordero F, Forde D, Ita H, Kosower D A and Maitre D 2008 One-Loop Calculations with BlackHat Nucl. Phys. Proc. Suppl. 183 313

63. Ellis R K, Giele W T, Kunszt Z, Melnikov K and Zanderighi G 2009 One-loop amplitudes for W + 3 jet production in hadron collisions JHEP 0901 012

64. Ossola G, Papadopoulos C G and Pittau R 2008 JHEP 0803 042 http://www.ugr.es/~pittau/CutTools/

65. Hirschi V, Frederix R, Frixione S, Garzelli M V, Maltoni F and Pittau R 2011 Automation of one-loop QCD corrections JHEP 1105 044

66. Cascioli F, Maierhoefer P and Pozzorini S 2012 Scattering Amplitudes with Open Loops Phys. Rev. Lett. 108 111601

67. Cullen G et al. 2010 Recent Progress in the Golem Project Nucl. Phys.

Proc. Suppl. 205-206 67

68. Oleari C 2010 The PowHeg box Nucl. Phys. Proc. Suppl. 205-206 34

69. Frixione S and Webber B R 2007 Matching NLO QCD computations with parton shower simulations: the POWHEG method, JHEP 0711 070

70. Alwall J et al. aMC@NLO http://amcatnlo.web.cern.ch/amcatnlo/

71. Bern Z, Ozeren K, Dixon L, Hoeche S, Febres Cordero F, Ita H, Kosower D and Maître D 2012 High multiplicity processes at NLO with BlackHat and Sherpa in Proceedings of Loops and Legs 2012 http://arxiv.org/pdf/1210.6684.pdf

72. Lopez-Val D, Goncalves-Netto D, Mawatari K, Plehn T, Wigmore I 2012 MadGolem: automating NLO calculations for New Physics http://arxiv.org/abs/1209.2797

73. Cullen G, Greiner N, Heinrich G, Luisoni G, Mastrolia P, Ossola G, Reiter T and Tramontano F 2012 Automated One-Loop Calculations with GoSam Eur. Phys. J., no. C 72 1889 http://gosam.hepforge.org/

74. Bevilacqua G, Czakon M, Garzelli M V, van Hameren A, Kardos A, Papadopoulos C G, Pittau R and Worek M 2011 HELAC-NLO Comp. Phys. Comm. 184 986

75. Lepage G P 1978 A New Algorithm for Adaptive Multidimensional Integration Journal of Computational Physics 27 192–203

76. VEGAS - GNU Scientific Library -- Reference Manual http://www.gnu.org/software/gsl/manual/html_node/VEGAS.html

77. Kawabata S 1995 A new version of the multi-dimensional integration and event Comp. Phys. Comm. 88 309-326

78. Yuasa F, Tobimatsu K and Kawabata S 2003 Parallelization of the multidimensional integration Nuclear Instruments and Methods in Physics Research A 502 599–601

79. Jadach S 2000 Foam: Multi-Dimensional General Purpose Monte Carlo Generator With SelfAdapting Simplical Grid Comp. Phys. Comm. 130 244-259

80. Ohl T 1999 VAMP, Version 1.0: Vegas AMPlified: Anisotropy, Multi-channel sampling and Parallelization http://whizard.hepforge.org/vamp.pdf

81. Nason P 2007 MINT: a Computer Program for Adaptive Monte Carlo Integration and Generation of Unweighted Distributions http://xxx.lanl.gov/abs/arXiv:0709.2085

82. de Doncker E et al. Parint: Parallel Integration Package http://www.cs.wmich.edu/parint/

83. Yuasa F, Ishikawa T, Kawabata S, Perret-Gallix D, Itakura K, Hotta Y and Okuda M 2000 Hybrid parallel computation of integration in GRACE in 6th International Workshop on New Computing Techniques in Physics Research (AIHEP99) http://arxiv.org/abs/hep-ph/0006268

84. Laccetti G, Lapegna M, Mele V, Romano D and Murli A 2012 A Double Adaptive Algorithm for Multidimensional Integration on Multicore Based HPC Systems International Journal of Parallel Programming 40 397-409 http://rd.springer.com/article/10.1007%2Fs10766-011-0191-4

85. Yuasa F, Ishikawa T, Hamaguchi N, Koike T and Nakasato N 2012 Acceleration of Feynman loop integrals in high-energy physics on many core GPUs in Conference on Computational Physics (Kobe), 2012 http://acpp-coll.in2p3.fr/cgibin/twiki.source/pub/ACPP/PresentationsNotes/ccp2012-poster-v2.pdf

86. Kanzaki J 2012 Monte Carlo integration on GPU Eur. Phys. J. C 71 1559

87. Kondo K 1991 Dynamical likelihood method for reconstruction of events with missing momentum. 2: Mass spectra for 2 → ; 2 processes J. Phys. Soc. Ja. 60 836–844

88. Artoisenet P, Lemaître V, Maltoni F and Mattelaer O 2010 Automation of the matrix element reweighting method JHEP 1012 068, https://cp3.irmp.ucl.ac.be/projects/madgraph/wiki/MadWeight

89. Harrington R, Optimization of Matrix element methods using GPUs in Workshop on Future Computing, University of Edinburgh, 2011 http://indico.cern.ch/getFile.py/access?contribId=10&sessionId=3&resId=0&materialId=slides& confId=141309

90. Brandt O, Top Quark mass mesurement at Tevatron 2010 Il Nuovo Cimiento 33 73-80

91. Campbell J M, Giele W T and Williams C 2012 Extending the matrix element method to next-toleading order FERMILAB-CONF-12-176-T, FERMILAB-CONF-12-176-T http://arxiv.org/pdf/1205.3434.pdf

92. CDF Collaboration 2012 Measurements of the Top-quark Mass and the tt-bar Cross Section in the Hadronic tau + Jets Decay Channel at sqrt(s)=1.96 TeV Phys. Rev. Lett. 109 192001

93. Fujimoto J, Ishikawa T and Perret-Gallix D 2005 High precision numerical computations: A case for an HAPPY design ACPP IRG note: ACPP-N-1 2005 http://emc2.in2p3.fr/cgibin/twiki.source/pub/ACAT/PresentationsNotes/Highprecisionnumericalcomputatio3.pdf

94. Lu M, He B and Luo Q, Supporting Extended Precision on Graphics

Processors in DaMoN '10 Proceedings of the Sixth International Workshop on Data Management on New Hardware, p 19- 26 http://www.ntu.edu.sg/home/bshe/precision_damon10.pdf

95. Jarp S, Lazzaro A, Nowak A and Valsan L 2012 Comparison of Software Technologies for Vectorization and Parallelization CERN openlab http://openlab.web.cern.ch/sites/openlab.web.cern.ch/files/technical_ documents/Evaluation_of_P arallel_Technology.pdf

96. Dong X, Cooperman G and Apostolakis J, Multithreaded Geant4: Semi-Automatic Transformation into Scalable Thread-Parallel Software College of Computer Science, Northeastern University, Boston and PH/SFT, CERN (Switzerland), http://citeseerx.ist.psu.edu/viewdoc/downlo ad?doi=10.1.1.174.8538&rep=rep1&type=pdf

97. Garren L and Lebrun P. StdHep: StdHep provides a common output format for Monte Carlo events http://cepa.fnal.gov/psm/stdhep/

98. Dobbs M and Hansen J B 2001 HepMC - a C++ Event Record for Monte Carlo Generators Comp. Phys. Comm. 134 41 https://savannah.cern.ch/projects/hepmc/

99. Binoth T, Boudjema F, Dissertori G, Lazopoulos A, Denner A, Dittmaier S, Frederix R, Greiner N, Hoche S, Giele W, Skands P, Winter J, Gleisberg T, Archibald J, Heinrich G, Krauss F, Maitre D, Huber M, Huston J, Kauer N, et al. 2010 A proposal for a standard interface between Monte Carlo tools and one-loop programs Comp. Phys. Comm. 181 1612

Chapter 2

STATISTICAL CORRELATIONS OF THE N-PARTICLE MOSHINSKY MODEL

Hsuan Tung Peng[1,2] and Yew Kam Ho[1]

[1]Institute of Atomic and Molecular Sciences, Academia Sinica, Taipei 10617, Taiwan

[2]Department of Physics, National Taiwan University, Taipei 10617, Taiwan

ABSTRACT

We study the correlation of the ground state of an N-particle Moshinsky model by computing the Shannon entropy in both position and momentum spaces. We have derived the Shannon entropy and mutual information with analytical forms of such an N-particle Moshinsky model, and this helps us test the entropic uncertainty principle. The Shannon entropy in position space decreases as interaction strength increases. However, Shannon entropy in momentum space has the opposite trend. Shannon entropy of the whole system satisfies the equality of entropic uncertainty principle. Our results also indicate that, independent of the sizes of the two subsystems, the mutual information increases monotonically as the interaction strength increases.

INTRODUCTION

Since understanding the correlation of quantum many body problems is crucial to quantum information processes, and quantum systems described as harmonically confined systems with tunable interaction parameters are promising for development in quantum information processes, such quantum systems hence provide us a motivation to study correlation in a solvable many body system—an N-particle Moshinsky model.

In the Moshinsky model [1], the system is confined in harmonic traps and inter-particle interaction also takes a harmonic form. The entropies in this system can be solved exactly, helping us investigate correlations and test the entropic uncertainty principle of two subsystems containing arbitrary numbers of particles. Several topics about the correlation of Moshinsky model, such as the statistical and quantum correlation of the two-electron Moshinsky model [2–4], three-electron Moshinsky model and applying a uniform magnetic field in the two-electron model [5], and the quantum correlation in N-particle Moshinsky model [6], have been studied recently. In our present work, we focus on three topics: understanding statistical correlations, testing the entropic uncertainty principle, and comparing the statistical correlation to the quantum correlation of an N-particle Moshinsky model.

For the first topic, to describe the statistical information in the system, we introduce Shannon entropy, which is a measure of uncertainty of the random variables. Usually, Shannon entropy is applied to describe delocalization or localization of the system. Using different phase-spaces to measure Shannon entropy will lead to different expressions. Studies of Shannon entropy in position space and momentum space of atomic systems have been carried out [7–14]. We calculate Shannon entropy in position basis and momentum basis to discuss the correlation. From information theory [15], mutual information [16] is a general measure of correlation of two subsystems, and it has been applied to study correlation in systems [17–19]. In an N-particle Moshinsky model, we discuss the correlation between two subsystems, one containing p particles and the other containing N-p particles. We can derive analytic wave function in position and momentum spaces, and from which Shannon entropy in both spaces can be calculated, leading to understanding of the relationship between these three factors, p, N, g (coupling coefficient of the Moshinsky model).

In order to discuss the statistical correlations of the system, we show the definition of the quantities we consider here. In [2], the definition of Shannon entropy, *one*-particle Shannon entropy and mutual information in position space is given, and we extend the definition to a system with N particles as follows:

$$S_{pos} = -\int dx_1...dx_N \, \Gamma(x_1,...,x_N) \ln \Gamma(x_1,...,x_N) \;,$$

$$S_{pos}^{(p)} = -\int dx_1...dx_p \, \gamma(x_1,...,x_p) \ln \gamma(x_1,...,x_p) \;,$$

$$I_{pos}^{(p,N-p)} = S_{pos}(\Gamma(x_1,...,x_N) \,\|\, \gamma(x_1,...,x_p)\gamma(x_{p+1},...,x_N)), \tag{1}$$

where x_i is the position of i-th particle. S_{pos} and S(p)pos are Shannon entropy of the whole system calculated by $\Gamma(x_1,..., x_N)$, the probability density function in position space, and the p-particle Shannon entropy in position space calculated by $\gamma(x_1,...,x_p) = \int dx_{p+1}...dx_N \Gamma(x_1,...,x_N)$, the reduced probability density function, respectively. $I_{pos}^{(p,N-p)}$ is mutual information of the composite system consists of two groups: a group with p particles and the other group $N-p$ particles, and it is defined by the relative entropy between the distribution $\Gamma(x_1,...,x_N)$ and $\gamma(x_1,...,x_p)\gamma(x_{p+1},...,x_N)$, and it can be calculated in a simple formula: $I_{pos}^{(p,N-p)} = S_{pos}^{(p)} + S_{pos}^{(N-p)} - S_{pos}$. Note that when $p=N$, the p-particle Shannon entropy is just the Shannon entropy of the whole system, i.e., $S_{pos}^{(N)} = S_{pos}$.

Also, all the counterpart quantities in momentum space can be defined as:

$$S_{mom} = -\int dq_1...dq_N \Lambda(q_1,...,q_N) \ln \Lambda(q_1,...,q_N),$$

$$S_{mom}^{(p)} = -\int dq_1...dq_p \lambda(q_1,...,q_p) \ln \lambda(q_1,...,q_p),$$

$$I_{mom}^{(p,N-p)} = S_{mom}(\Lambda(q_1,...,q_N) \| \lambda(q_1,...,q_p)\lambda(q_{p+1},...,q_N)), \qquad (2)$$

where q_i is momentum of the i-th particle. S_{mom} and $S^{(p)}_{mom}$ are Shannon entropy of the whole system calculated by $\Lambda(q_1,...,q_2)$, the probability density function in momentum space, and the p-particle Shannon entropy in momentum space calculated by $\lambda(q_1,...,q_p) = \int dq_{p+1}...dq_N \Lambda(q_1,...,q_N)$, the reduced probability density function, respectively. $I_{mom}^{(p,N-p)}$ is the mutual information in momentum space, and it can be calculated in a simpler formula: $I_{mom}^{(p,N-p)} = S_{mom}^{(p)} + S_{mom}^{(N-p)} - S_{mom}$. For the second topic, the entropic uncertainty principle has been investigated in [20], and entropic uncertainty relations in atomic systems were discussed in some studies [2,21]. In this model, by calculating Shannon entropies in position and momentum space, we can test the entropic uncertainty principle [20], as:

$$S_{pos}^{(p)} + S_{mom}^{(p)} \geq p(1+\ln \pi), \qquad S_{pos} + S_{mom} \geq N(1+\ln \pi). \qquad (3)$$

From the entropic uncertainty principle, the sum of entropy in phase-spaces, $S_{pos} + S_{mom}$, can be considered as the entropy of a product distribution in phase-spaces, $\Gamma(x_1,...,x_N)\Lambda(q_1,...,q_N)$, and such a way of thinking is also valid for the case of reduced distribution.

For the third topic, comparing to Shannon entropy, von Neumann entropy is a measure of quantum information and is widely used in many atomic systems [22–27]. For a bipartite pure state, von Neumann entropy is half of the quantum mutual information [28]; therefore it can also be a good measure of quantum correlation. The eigenvalue structure of N-particle Moshinsky model has been studied in [29], and results of von Neumann entropy have been given by [6]. Shannon entropy does not equal to von Neumann entropy most of time.

However, we can compare the behavior of statistical and quantum correlation to three factors p, N, g. In this article, we discuss the first topic in Sections 3.1 and 3.2, the second topic in Section 3.3, and the third topic in Section 3.4.

MOSHINSKY MODEL

For the N-particle Moshinsky model, the Hamiltonian is:

$$H = \sum_{i=1}^{N} (-\frac{\hbar^2}{2m} \frac{d^2}{dx_i^2} + \frac{1}{2} m\omega^2 x_i^2) + \sum_{1 \leq i < j \leq N} \kappa\omega^2 (x_i - x_j)^2 .$$
(4)

Take the scaled unit $(x \to \sqrt{\frac{\hbar}{m\omega}} x, E \to \hbar\omega E)$, and let $g = \frac{\kappa}{m}$. The Hamiltonian of Equation (4) turns into:

$$H = \sum_{i=1}^{N} (-\frac{1}{2} \frac{d^2}{dx_i^2} + \frac{1}{2} x_i^2) + \sum_{1 \leq i < j \leq N} g(x_i - x_j)^2.$$
(5)

In order to solve the wave function of this system, we transform the original coordinates into Jacobi coordinates:

$$\begin{cases} X = \dfrac{x_1 + ... + x_N}{\sqrt{N}} , \\ X_i = \sqrt{\dfrac{i-1}{i}}(x_i - \dfrac{1}{i-1}\sum_{k=1}^{i-1} x_k), \quad i = 2,3,...,N . \end{cases}$$
(6)

By using such transformation, we can separate the Hamiltonian in Jacobi coordinates:

$$H = (-\frac{1}{2} \frac{d^2}{dX^2} + \frac{1}{2} X^2) + \sum_{i=2}^{N} (-\frac{1}{2} \frac{d^2}{dX_i^2} + \frac{1}{2}\Omega^2 X_i^2) = H_X + \sum_{i=2}^{N} H_{X_i} .$$
(7)

Here $\Omega = \sqrt{1+2Ng}$, and now we can derive the exact wave function for this system:

$$\psi_{(n_{CM}, n_2,...,n_N)}(X, X_2,...,X_N) = (\frac{1}{\sqrt{2^{n_{CM}} n_{CM}!}})(\frac{1}{\pi})^{\frac{1}{4}} e^{-X^2/2} H_{n_{CM}}(X) \prod_{i=2}^{N} (\frac{1}{\sqrt{2^{n_i} n_i!}})(\frac{\Omega}{\pi})^{\frac{1}{4}} e^{-\frac{\Omega}{2}\sum_{i=2}^{N} X_i^2} H_{n_i}(\sqrt{\Omega}X_i),$$
(8)

where the quantum numbers $(n_{CM}, n_2,...,n_N)$ label the state in position space, and all the quantum numbers must be non-negative integer, and $H_n(x)$ is the Hermite polynomial.

On the other hand, to derive the wave function in momentum space, we can apply the Fourier transform to Equation (8). However, a simpler way to derive such wave function is to rewrite the Hamiltonian in momentum coordinates, which is given by:

$$H = \frac{1}{2}Q^2 - \frac{1}{2}\frac{d^2}{dQ^2} + \sum_{i=2}^{N}(\frac{1}{2}Q_i^2 - \frac{1}{2}\Omega^2\frac{d^2}{dQ_i^2}).$$

$$(9)$$

In Equation (9), we use the Jacobi basis in momentum space, which is

$$Q = \frac{q_1 + q_2 + \dots + q_n}{\sqrt{N}}, Q_i = \sqrt{\frac{i-1}{i}}(q_i - \frac{1}{i-1}\sum_{k=1}^{i-1}q_k)$$
. Thus, the wave function in momentum

space is:

$$\psi_{(m_{CM},m_2,\dots,m_N)}(Q,Q_2,\dots,Q_N) = (\frac{1}{\sqrt{2^{m_{CM}}m_{CM}!}})(\frac{1}{\pi})^{\frac{1}{4}}e^{-Q^2/2}H_{m_{CM}}(Q)\prod_{i=2}^{N}(\frac{1}{\sqrt{2^{m_i}m_i!}})(\frac{1}{\pi\Omega})^{\frac{1}{4}}e^{-\frac{1}{2\Omega}\sum_{i=2}^{N}Q_i^2}H_{n_i}(\frac{Q_i}{\sqrt{\Omega}}),$$

$$(10)$$

where the quantum numbers $(m_{CM}, m_2, \dots, m_N)$ label the state in momentum space, and all the quantum numbers must be non-negative integer. By obtaining the wave function in position space and momentum space we can calculate Shannon entropies in both spaces.

SHANNON ENTROPY AND TESTING ENTROPIC UN-CERTAINTY PRINCIPLE

Position Space

The ground state wave function of position space in Jacobi coordinate is:

$$\psi_{(0,0,\dots,0)}(X, X_2, \dots, X_N) = (\frac{1}{\pi})^{\frac{1}{4}}e^{-X^2/2}(\frac{\Omega}{\pi})^{\frac{N-1}{4}}e^{-\frac{\Omega}{2}\sum_{i=2}^{N}X_i^2}$$

$$(11)$$

To calculate p-particle Shannon entropy, we first construct the p-particle reduced probability density function, as:

$$\gamma(x_1,x_2,\dots,x_p) = C_p \exp[-\frac{1+\Omega(N-1)}{N}\sum_{1\le i\le p}x_i^2 + \frac{2(\Omega-1)}{N}\sum_{1\le i\le j\le p}x_ix_j + \frac{(N-p)(\Omega-1)^2}{N(N-p+p\Omega)}(\sum_{1\le i\le p}x_i)^2],$$

$$(12)$$

where $C_p = \pi^{-\frac{p}{2}}\Omega^{\frac{p}{2}}[\frac{N-p}{N} + \frac{p}{N}\Omega]^{-1/2}$.

Next, we calculate the p-particle Shannon entropy mentioned in the second line of Equation (1) with some special treatment. Let the Jacobi coordinate of p-particle system be $Y = \frac{1}{\sqrt{p}}\sum_{i=1}^{p}x_i, y_i = \sqrt{\frac{i-1}{i}}(x_i - \frac{1}{i-1}\sum_{k=1}^{i-1}x_k)$, and define the following three quantities:

$$r^2 = \sum_{i=2}^{p} y_i^2, \; a = \sum_{i=1}^{p} x_i^2, \; \text{and} \; b = \sum_{1 \le i \le j \le N} x_i x_j \;.$$
$$\tag{13}$$

Thus, we can write down the relation between Y, r and a, b, as:

$$\begin{cases} Y^2 = \dfrac{1}{p}a + \dfrac{2}{p}b \\ r^2 = \dfrac{p-1}{p}a - \dfrac{2}{p}b \end{cases} \Rightarrow \begin{cases} a = Y^2 + r^2 \\ b = \dfrac{(p-1)Y^2 - r^2}{2} \end{cases} .$$
$$\tag{14}$$

Since $\gamma(x_1, x_2, \dots, x_p) = C_p \exp[-\dfrac{1+\Omega(N-1)}{N}a + \dfrac{2(\Omega-1)}{N}b + \dfrac{(N-p)(\Omega-1)^2}{N(N-p+p\Omega)}(a+2b)]$, now we can put (14) into the reduced probability density function and obtain:

$$\gamma(Y, y_2, \dots, y_p) = C_p \exp[-\dfrac{\Omega N}{N-p+p\Omega}Y^2 - \Omega r^2] = C_p \exp[-C_1 Y^2 - C_2 r^2],$$
$$\tag{15}$$

where $C_1 = \dfrac{\Omega N}{N-p+p\Omega}, C_2 = \Omega$.

We then calculate the p-particle Shannon entropy in $Y, y_2, \dots, y_p$ coordinates (the determinant of the Jacobian is 1), and the result becomes:

$$S_{pos}^{(p)} = -\int dY dy_2 \dots dy_p \gamma(Y, \dots, y_p) \ln \gamma(Y, \dots, y_p) = C_p \sqrt{\dfrac{\pi}{C_1}} (\sqrt{\dfrac{\pi}{C_2}})^{p-1} (\dfrac{p}{2} - \ln C_p) = \dfrac{p}{2} - \ln C_p.$$
$$\tag{16}$$

Take $p = N$ in Equation (16), we obtain Shannon entropy of the whole system in position space. Now, the mutual information in position space can be computed as follows:

$$I_{pos}^{(p, N-p)} = S_{pos}^{(p)} + S_{pos}^{(N-p)} - S_{pos} = -\dfrac{1}{2}\ln\Omega + \dfrac{1}{2}\ln[\dfrac{N^2\Omega + (N-p)p(1-\Omega)^2}{N^2}]$$
$$\tag{17}$$

Next, we plot the results for these quantities. For Shannon entropy of the whole system, results are shown in Figure 1. In the weak interaction region (when g is small, which means there is almost no interaction), we observe that the higher number to the total particles, the higher value is to the Shannon entropy; however, in the strong interaction region (when g is large), the trend is opposite. This fact can be deduced from Equation (16) by taking $p = N$, with:

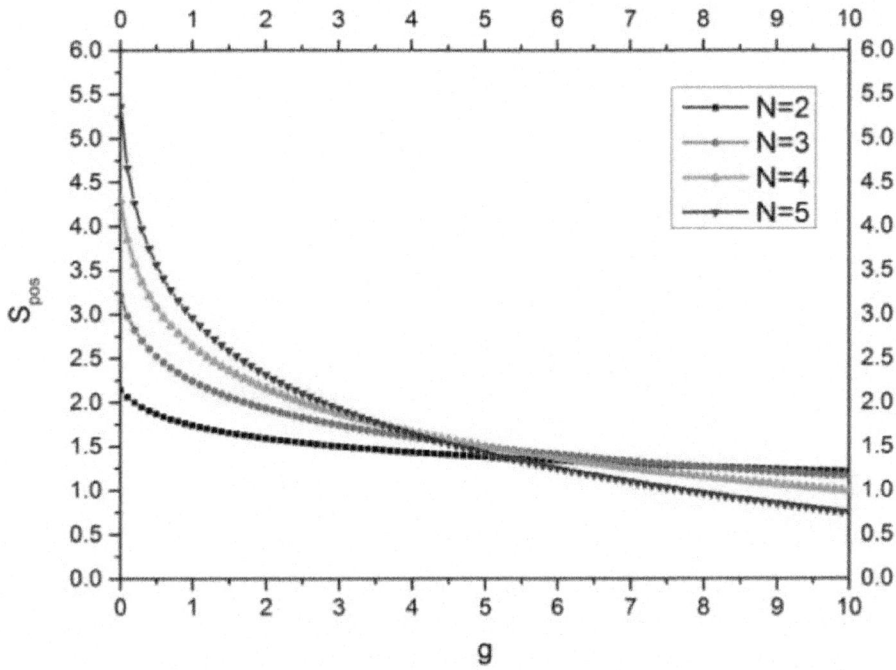

Figure 1: Shannon entropy of whole system for $N = 2$ to 5.

$$S_{pos} = \frac{N}{2} - \ln C_N = \frac{N}{2}(1 + \ln \pi) - \frac{N-1}{2} \ln \sqrt{1 + 2Ng}$$

$$\frac{dS_{pos}}{dg} = -\frac{N(N-1)}{2} \frac{1}{1+2Ng}$$

(18)

Therefore, for small g, Shannon entropy is proportional to N, and the derivative of Shannon entropy to the interaction strength g is always negative. The effect of large N is that the decreasing rate is greater, and the distribution is more sensitive to localization when interaction increases.

If we consider p-particle Shannon entropy and fix the total number of particles, the p-particle Shannon entropy would decrease as interaction getting stronger. The larger p number to the p-particle system, the faster decreasing rate is to the p-particle Shannon entropy. For different N, the trend is similar, so we just show results for a system with $N = 8$ particles in Figure 2, as an example.

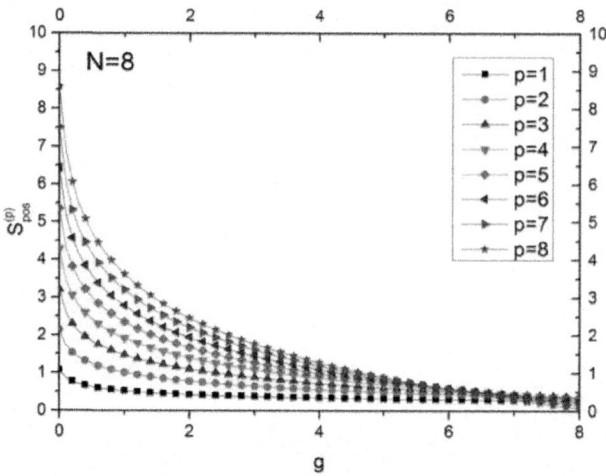

Figure 2: p-particle Shannon entropy for fixed $N=8$, $p=1$ to 8.

For mutual information, when p is closer to $N-p$ the mutual information becomes greater, and the maximum value occurs at $p=\lfloor N2 \rfloor$. The difference in total interaction between two subsystems is a result of different partitions for such two subsystems. When the size of two subsystems is getting closer, the total interaction between two subsystems is stronger.

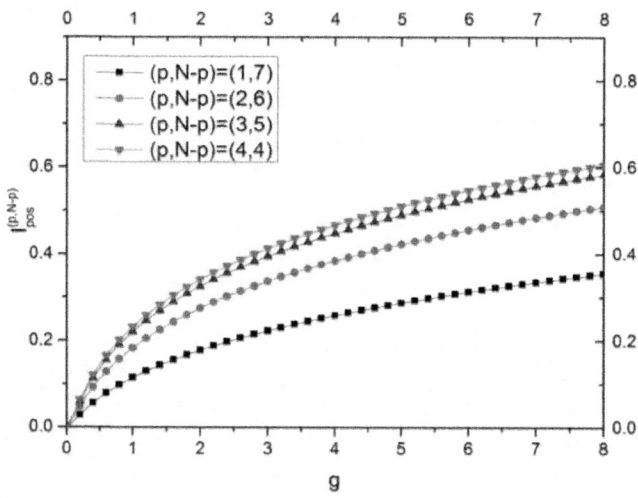

Figure 3: Mutual information in position space for total particles number $N = 8$, the black line labeled as square is for partition of $(p, N-p) = (1,7)$, and the pink line labeled as down triangle is for partition $(p, N-p) = (4,4)$, which are the lowest and highest mutual information, respectively, in position space.

As a result, when p is closer to N-p the mutual information becomes greater. The Shannon entropy is decreasing as the interaction is getting stronger; however, mutual information is increasing as the interaction strength increases. The mutual information can be considered as the correlation between two subsystems (one part contains p particles, and the other contains N-p particles), so this fact shows that when the interaction strength increases, the correlation of these two subsystems becomes stronger. Figure 3 shows the mutual information when $N = 8$.

Momentum Space

We can compute Shannon entropy in momentum space in a similar manner. The ground state wave function of momentum space in Jacobi coordinates is:

$$\phi_{(0,0,...,0)}(Q,Q_2,...,Q_N) = (\frac{1}{\pi})^{\frac{1}{4}} e^{-Q^2/2} (\frac{1}{\pi\Omega})^{\frac{N-1}{4}} e^{-\frac{q^2}{2\Omega}}.$$

(19)

Following the same procedure as that in the position space, we calculate Shannon entropy for whole system and p-particle Shannon entropy respectively. The p-particle Shannon entropy is:

$$S_{mom}^{(p)} = D_p \sqrt{\frac{\pi}{D_1}} (\sqrt{\frac{\pi}{D_2}})^{p-1} (\frac{p}{2} - \ln D_p) = \frac{p}{2} - \ln D_p,$$

(20)

where $D_p = (\frac{1}{\pi\Omega})^{p/2} (\frac{N-p}{N} + \frac{p}{N\Omega})^{-1/2}, D_1 = \frac{N}{\Omega(N-p)+p}, D_2 = \frac{1}{\Omega}$. By taking $p=N$ in Equation (20), Shannon entropy for whole system can be derived.

The mutual information can be derived as well:

$$I_{mom}^{(p,N-p)} = S_{mom}^{(p)} + S_{mom}^{(N-p)} - S_{mom} = -\frac{1}{2}\ln\Omega + \frac{1}{2}\ln[\frac{(N-p)p(\Omega^2+1)+((N-p)^2+p^2)\Omega}{N^2}] = I_{pos}^{(p,N-p)}.$$

(21)

Shannon entropy of the whole system in momentum space is shown in Figure 4. Unlike in position space, the Shannon entropy increases as interaction strength increases. This means that the momentum space distribution is delocalized as interaction increases. The fact can be derived from Equation (20):

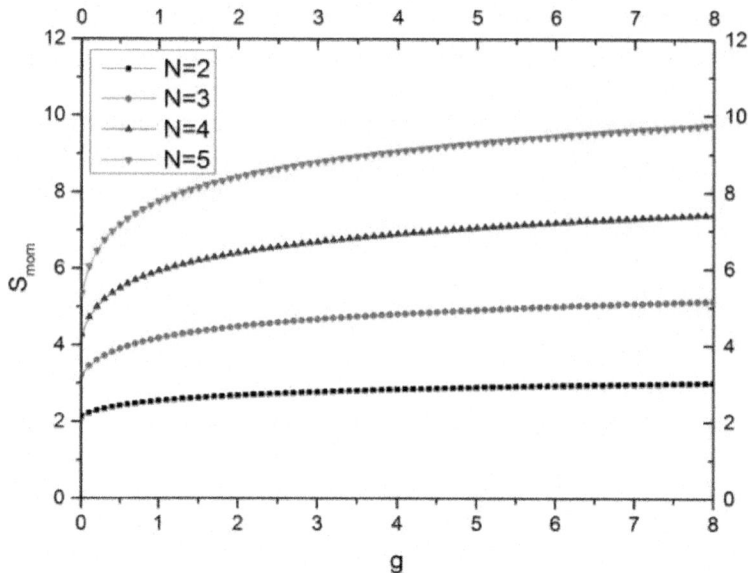

Figure 4: Shannon entropy of the whole system in momentum space, from $N=2$ to 5.

$$S_{mom} = \frac{N}{2} - \ln D_N = \frac{N}{2}(1+\ln\pi) + \frac{N-1}{2}\ln\sqrt{1+2Ng}$$

$$\frac{dS_{mom}}{dg} = \frac{N(N-1)}{2}\frac{1}{1+2Ng}$$

(22)

Since the derivative of Shannon entropy to the interaction strength is positive, Shannon entropy in momentum space is increasing as interaction strength increases. Furthermore, when N is a larger number the rate of increase of the Shannon entropy becomes greater, and the delocalization is more sensitive to the interaction strength.

The p-particle Shannon entropy is increasing as interaction strength increases, and for a given interaction strength, a system with larger p value has larger Shannon entropy as compared to a system with smaller p vale. Here, we show results in Figure 5 for a fixed total number of particles $N = 8$, as an example.

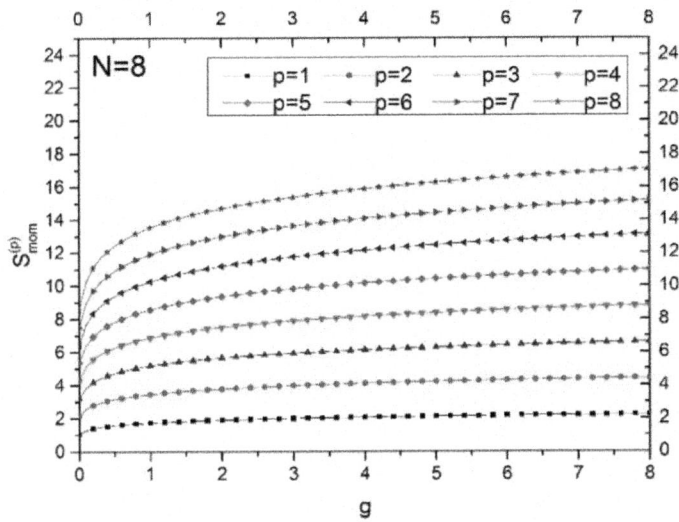

Figure 5: p-particle Shannon entropy for fixed $N = 8$, and $p = 1$ to 8.

Next from our results, we discuss the trends for Shannon entropy in different spaces. When the interaction is getting stronger, Shannon entropy in momentum space is increasing, while in position space Shannon entropy is decreasing. An explanation of the opposite trend for Shannon entropy in different spaces is that when interaction becomes stronger, particles tend to become closer in position space (concentrate to some small region); therefore, the uncertainty in position space decreases, which also means that Shannon entropy will decrease in position space. However, in momentum space, when interaction becomes stronger the motion of particles is more chaotic, which means the uncertainty is greater and Shannon entropy is increasing in momentum space. Moreover, by taking suitable parameters, we can reduce our results to the case of total number of particles $N = 2$, and they are the same as those in [2].

For mutual information in momentum space, the results are exactly the same as those in position space. Although the behaviors to the interaction are different in position and momentum spaces, the mutual information is the same, independent of which phase space we have chosen. It is further observed that mutual information increases as interaction strength is increased.

In Figure 6, we show results for a system with $N = 8$ in momentum space, indicating that such results in momentum space are exactly the same as those in Figure 3 for position space.

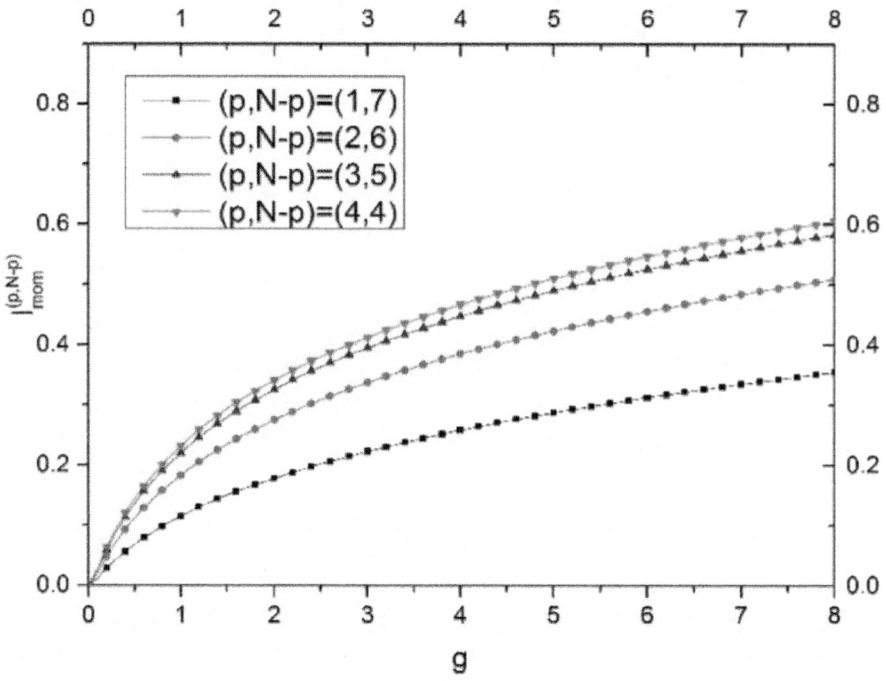

Figure 6: Mutual information in momentum space for total particles number $N = 8$, the black line labeled as square is for partition of $(p, N\text{-}p) = (1,7)$, and the pink line labeled as down triangle is for partition $(p, N\text{-}p) = (4,4)$ which are the lowest and highest mutual information, respectively, in momentum space.

Relation of Two Spaces and Testing Entropic Uncertainty Principle

The relationship between Shannon entropies in two spaces can be shown in Figure 7. For $g \geq 0$, Shannon entropy in momentum space is greater than or equal to the Shannon entropy in position space. They are equal to each other only when $g = 0$. The sum of these two Shannon entropies is a constant only when $p = N$ (the whole system Shannon entropy), and for all other cases, we observe that the entropy sum increases for increasing interaction strength.

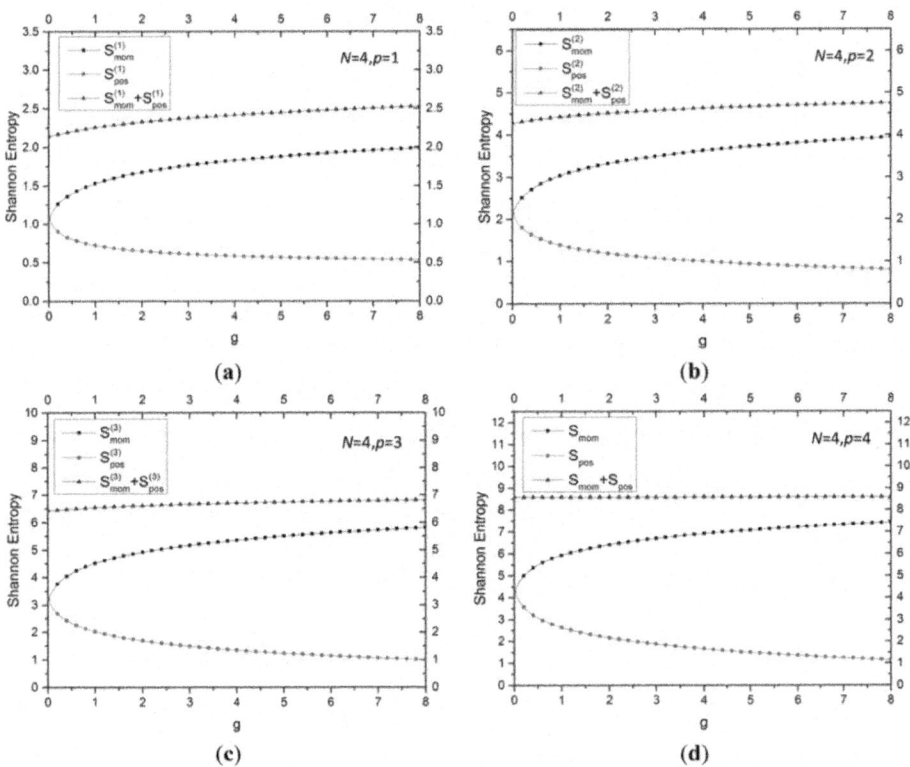

Figure 7: All the four figures are the case of total number of particles $N = 4$, and **(a)**, **(b)**, **(c)**, **(d)** are the cases of $p = 1, p = 2, p = 3, p = 4$, respectively. The red line labeled as circle is for Shannon entropy in position space, the black line labeled as square is for Shannon entropy in momentum space, and the blue line labeled as up triangle is for the sum of these two quantities.

According to the entropic uncertainty principle [20], inequality (3) should be satisfied. We now can show our results indeed satisfy these two inequalities. From Equations (18) and (22), we have obtained $S_{pos} + S_{mom} = N - \ln(C_N D_N) = N + N \ln \pi$, which is the equality of the uncertainty principle. Furthermore, we find that the sum of Shannon entropy in position and momentum space is independent of interaction strength. On the other hand, from Equations (16) and (20), we can prove that:

$$S_{pos}^{(p)} + S_{mom}^{(p)} = p - \ln(C_p D_p)$$

$$= p - \ln\{(\frac{1}{\pi})^p [(\frac{N-p}{N} + \frac{p}{N\Omega})(\frac{N-p}{N} + \frac{p}{N}\Omega)]^{-1/2}\}$$

$$= p + p\ln\pi + \frac{1}{2}\ln[1 + \frac{(N-p)p}{N^2}(\sqrt{\Omega} - \frac{1}{\sqrt{\Omega}})^2] \geq p(1 + \ln\pi)$$

$$(23)$$

The last inequality holds because the argument in ln is greater than 1, so this term is greater than zero, and the equality is satisfied only when $p = N$ or $p = 0$, or there is no interaction. Also, from Equation (23) it is quite straightforward to see that the sum of p-particle Shannon entropy in these two spaces increases monotonically when the interaction strength is increased.

Comparing Statistical Correlation to Quantum Correlation

As mentioned in the Introduction, quantum mutual information is twice the von Neumann entropy. The results of von Neumann entropy of the N-particle Moshinsky model are given in [6], thus we can compare classical mutual information with quantum mutual information. Mutual information of position space is the same as in momentum space, so we denote them as $I_c^{(p,N-p)}$, and quantum mutual information is denoted by $I_\varrho^{(p,N-p)}$, where $(p, N-p)$ indicates the partition of these two subsystems. In Figure 8, we show the comparison of classical and quantum mutual information with total number of particles $N = 8$, as an example.

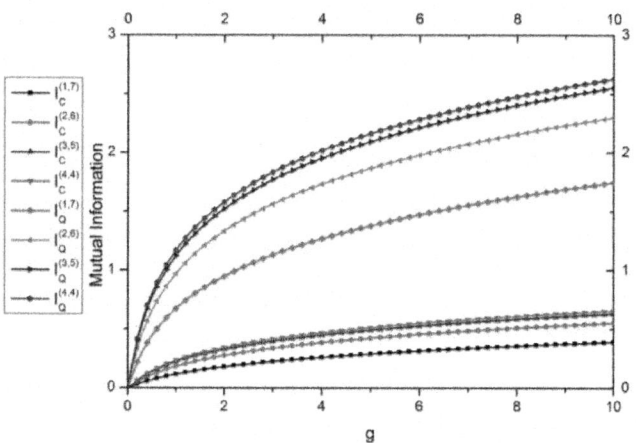

Figure 8: The comparison of classical and quantum mutual information with total number of particles $N = 8$. The lower four lines are classical mutual information, and the upper four lines are quantum mutual information.

For other N, the trends are similar. Classical and quantum mutual information share the same two trends: one trend is that both mutual information are monotonically growing up when the interaction strength increases, and the other trend is that for a given interaction strength, the closer p to $N-p$, the greater value is to the mutual information. On the other hand, quantum mutual information is growing faster than classical mutual information. This implies that quantum correlation is more sensitive to interaction strength than statistical correlation.

SUMMARY AND CONCLUSIONS

In the present work, we have analytically derived the Shannon entropy in position and momentum spaces for the ground state of an N-particle Moshinsky model. We show results of these entropies for total number of particles N and arbitrary number of particles p in subsystems, and apply such results to three topics: discussing the statistical correlations, testing the entropic uncertainty principle, and comparing the classical mutual information to quantum mutual information for this model.

For the first topic, we have observed that the behaviors of Shannon entropy are different in different phase-spaces; however, the mutual information is the same for both spaces. When the interaction is getting stronger, the Shannon entropy in momentum space is increasing (delocalization), while in position space Shannon entropy is decreasing (localization). Moreover, Shannon entropy is dependent of N and p. The rate of change of the Shannon entropy is larger when N or p is a larger number. In momentum space the rate of change is positive, while in position space the rate of change is negative. The Shannon entropies in both spaces are the same if there is no interaction. When the number p or N is getting larger, the Shannon entropy becomes greater when there is no interaction. Using mutual information as a measurement of correlation, the statistical correlation is the same for both spaces, which implies that correlation in the ground state of an N-particle Moshinsky model is independent of the spaces we have chosen. Mutual information depends on the interaction between particles and the partition of two subsystems, and it increases monotonically when the interaction strength is increased. Furthermore, the mutual information gets a larger value when the numbers of particles in the two subsystems are closer to each other. The maximum mutual information occurs when the sizes of the two subsystems are the same, *i.e.*, $p = N\text{-}p$.

For the second topic, Shannon entropies of the whole system and p-particle Shannon entropy satisfy inequality (3), the entropic uncertainty principle. The Shannon entropy of the whole system always satisfies the equality of entropic uncertainty principle whatever the interaction strength is, while p-particle Shannon entropy only satisfies the equality when interaction is zero. Furthermore, the sum of p-particle Shannon entropy in both spaces is increasing as interaction strength increases, while the sum of Shannon entropy of the whole system remains constant, and is independent of interaction strength. For the third topic, we show that classical and quantum mutual information have similar behaviors to the interaction strength and to the partition of subsystems. Both monotonically increase for increasing interaction strength, and when the sizes of the two subsystems are closer to each other the mutual information would have a larger value. The increasing rate of quantum mutual information is greater than that of classical mutual information.

ACKNOWLEDGMENTS

This work was supported by the Ministry of Science and Technology in Taiwan. We are thankful to the anonymous referees for their valuable suggestions. PACS code: 03.65. –w; 03.67. –a

AUTHOR CONTRIBUTIONS

Yew Kam Ho set the research direction and managed the overall progress of this project. Hsuan Tung Peng conceived the present theoretical frame, and carried out numerical calculations for the results presented in the present work. Both authors contributed to analyzing the numerical data and provided physical interpretation of the present results. Both authors contributed to writing the paper, have read and approved the final manuscript.

REFERENCES

1. Moshinsky, H.M. How good is the Hartree-Fock approximation. *Am. J. Phys.* **1968**, *36*, 52–53.

2. Laguna, H.G.; Sagar, R.P. Statistical correlations in the Moshinsky atom. *Phys. Rev. A* **2011**, *84*, 012502.

3. Yanez, R.J.; Plastino, A.R.; Dehesa, J.S. Quantum entanglement in a soluble two-electron model atom. *Eur. Phys. J. D* **2010**, *56*, 141–150.

4. Manzano, D.; Plastino, A.R.; Dehesa, J.S.; Koga, T. Quantum entanglement in two-electron atomic models. *J. Phys. A Math. Theor.* **2010**, *43*, 275301.

5. Bouvrie, P.A.; Majtey, A.P.; Plastino, A.R.; Sanchez-Moreno, P.; Dehesa, J.S. Quantum entanglement in exactly soluble atomic models: the Moshinsky model with three electrons, and with two electrons in a uniform magnetic field. *Eur. Phys. J. D* **2012**, *66*.

6. Kościk, P.; Okopińska, A. Correlation effects in the Moshinsky model. *Few-Body Syst* **2013**, *54*, 1637–1640.

7. Laguna, H.G.; Sagar, R.P. Indistinguishability and correlation in model systems. *J. Phys. A Math. Theor.* **2011**, *44*, 185302.

8. Laguna, H.G.; Sagar, R.P. Phase-space position-momentum correlation and potentials. *Entropy* **2013**, *15*, 1516–1527.

9.   Laguna, H.G.; Sagar, R.P. Position–momentum correlations in the Moshinsky atom. *J. Phys. A Math. Theor.* **2012**, *45*, 025307.

10. Laguna, H.G.; Sagar, R.P. Wave function symmetry, symmetry holes, interaction and statistical correlation in the Moshinsky atom. *Physica A* **2014**, *396*, 267–279.

11. Guevara, N.L.; Sagar, R.P.; Esquivel, R.O. Shannon-information entropy sum as a correlation measure in atomic systems. *Phys. Rev. A* **2003**, *67*, 012507.

12. Shi, Q.; Kais, S. Finite Size Scaling for the atomic Shannon-information entropy. *J. Chem. Phys.* **2004**, *121*, 5611–5617.

13. Sen, K.D. Characteristic features of Shannon information entropy of confined atoms. *J. Chem. Phys.* **2005**,*123*, 074110.

14. Chatzisavvas, K.C.; Moustakidis, C.C.; Panos, C.P. Information entropy, information distances, and complexity in atoms. *J. Chem. Phys.* **2005**, *123*, 174111.

15. Shannon, C.E. A mathematical theory of communication. *Bell Syst. Tech. J.* **1948**, *27*, 379–423.

16. Cover, T.M.; Thomas, J.A. *Elements of Information Theory*, 2nd ed.; Wiley: Hoboken, NJ, USA, 1991; pp. 12–49.

17. Sagar, R.P.; Guevara, N.L. Mutual information and correlation measures in atomic systems. *J. Chem. Phys.* **2005**, *123*, 044108.

18. Sagar, R.P.; Guevara, N.L. Mutual information and electron correlation in momentum space. *J. Chem. Phys.* **2006**, *124*, 134101.

19. Sagar, R.P.; Laguna, H.G.; Guevara, N.L. Conditional entropies and position-momentum correlations in atomic systems. *Mol. Phys.* **2009**, *107*, 2071–2080.

20. Bialynicki-Birula, I.; Mycielski, J. Uncertainty relations for information entropy in wave mechanics.*Commun. Math. Phys.* **1975**, *44*, 129–132.

21. Guevara, N.L.; Sagar, R.P.; Esquivel, R.O. Information uncertainty-type inequalities in atomic systems. *J. Chem. Phys.* **2003**, *119*, 7030–7036.

22. Lin, C.H.; Ho, Y.K. Quantification of entanglement entropy in helium by the Schmidt–Slater decomposition method. *Few-Body Syst.* **2014**, *55*, 1141–1149.

23. Lin, C.H.; Ho, Y.K. Calculation of von Neumann entropy for hydrogen and positronium negative ions.*Phys. Lett. A* **2014**, *378*, 2861–2865.

24. Lin, C.H.; Lin, Y.C.; Ho, Y.K. Quantification of linear entropy for quantum entanglement in He, H⁻ and Ps⁻ions using highly-correlated Hylleraas functions. *Few-Body Syst.* **2013**, *54*, 2147–2153.

25. Lin, Y.C.; Ho, Y.K. Quantum entanglement for two electrons in the excited states of helium-like systems.*Can. J. Phys.* **2015**.

26. Lin, Y.C.; Lin, C.Y.; Ho, Y.K. Spatial entanglement in two-electron atomic systems. *Phys. Rev. A* **2013**, *87*, 022316.

27. Majtey, A.P.; Plastino, A.R.; Dehesa, J.S. The relationship between entanglement, energy and level degeneracy in two-electron systems. *J. Phys. A Math. Theor.* **2012**, *45*, 115309.

28. Jaeger, G. *Quantum Information—An Overview*, 1st ed.; Springer: New York, NY, USA, 2006; Chapter 5; pp. 85–86.

29. Pruski, S.; Maćkowiak, J.; Missuno, O. Reduced density matrices of a system of N coupled oscillators 3. Eigenstructure of the p-particle matric for the ground-state. *Rep. Math. Phys.* **1972**, *3*, 241–246.

Chapter 3

A NOVEL ELEMENTARY PARTICLE THEORY BASED ON EXTERNAL ENERGY ABSORBED AND RE-EMITTED BY ATOMS

Hans W. Giertz

GiertzTech AB, Gnesta, Sweden

ABSTRACT

In this study, all energy in the universe, here called energy quanta, originates from a singularity at the centre of the universe. The electron and the atom are completely passive; they absorb energy quanta by means of forced damped oscillators and the absorbed energy quanta are then re-emitted. Absorbed and stored energy quanta result in nuclear energy. Re-emitted energy quanta are fields. Re-emitted energy quanta operating on oscillators, in adjacent particles and atoms, result in forces: strong force, Coulomb force and gravitational force. Using this model may enable unification of elementary particle physics, general relativity, electromagnetic theory and quantum physics into one comprehensive theory.

INTRODUCTION

Today's elementary particle physics, quantum physics, electromagnetic theory and general relativity build on the assumption that electron and atomic forces, fields and energy are created or generated by the electron and atom itself without external influence. In the present study today's model is replaced by a fundamentally different model. In the new model all energy in the universe, here called energy quanta, originates from a singularity at the centre of the universe. The electron and the atom are completely passive; they absorb energy quanta by means of forced damped oscillators and the absorbed energy quanta are then re-emitted. Parts of this novel theory have been documented in eight different studies [1] -[8] .

The Higgs Field is an energy field that exists everywhere in the universe and is accompanied by a fundamental particle called the Higgs Boson [9] - [11] . It is proposed here that the Higgs Boson corresponds to one of the above energy quanta and that the Higgs Field corresponds to the matching energy quanta flow density.

Energy radiated from the singularity has been thoroughly measured and documented, including description of measurement technique [1] [2] and instrument technique [3] . It is a relatively simple task to repeat these measurements and to verify the existence of the singularity: its type of energy (energy quanta), its frequency spectrum and the direction towards the singularity. The direction to the singularity is towards north and along the earth's rotational axis [1] [2] .

Atoms absorb energy quanta at frequencies specific to the atom. Thus, the atom contains (forced damped) oscillators with natural frequencies equal to the frequencies of energy quanta generated by the singularity. Each oscillator can be described as an oscillator quantum. The absorbed energy quanta result in nuclear forces and stored nuclear energy [1] .

Electrons absorb energy quanta at a frequency specific to the electron. Thus, the electron contains an (forced damped) oscillator quantum with natural frequency equal to the frequency of energy quanta generated by the singularity. The absorbed energy quanta are called electron charge in today's physics. The absorbed energy quanta are then re-emitted uniformly into space by the oscillator quantum. Re-emitted energy quanta accounts for what is called Coulomb force, current, electric field, displacement and magnetic field in today's physics [12].

Atoms and electrons also absorb energy quanta at frequencies specific to gravity [2] . The absorbed energy quanta are re-emitted uniformly into space. The flow of energy quanta between oscillators in particles or masses results in mutual force of attraction, i.e. gravity [13] . It explains gravitational mass [5] [14] .

Particle, e.g. electron excess energy hv is superpositioned on energy quanta absorbed and re-emitted by oscillator quanta in the electron and the atomic nucleus [4] . Hence, the photon [15] [16] comprises superpositioned energy hv on energy quanta, originating from the singularity. The conversion is made in the oscillator quantum [4] . It also explains inertial mass [5] [17] [18] .

Parts of the present model have been bench marked on phenomena observed in astrophysics; it displays that it can replace general relativity [19] . The model provides simple solutions to metric expansion of space [6] , gravitational lensing [7] , gravitational redshift [8] and the black hole phenomena [8] .

The present model is mapped on quantum physics and demonstrates solutions to intrinsic angular momentum, intrinsic magnetic dipole momentum and electron wave-particle duality [20] [21] . A solution to the Pauli Exclusion Principle is presented using injection locking [22] [23] .

The purpose with the present paper is to provide a common theory for elementary particle physics, electromagnetic theory, quantum physics and astrophysics, i.e. a step towards the Theory of Everything (TOE). It complements and replaces parts of the Standard Model [24] , electromagnetic theory [12] , quantum physics [20] [21] , general relativity [19] and string theory [25] .

THEORETICAL MODEL

General

Today physics is described by different models: the Standard Model [24] , electromagnetic theory [12] , quantum physics [20] [21] , general relativity [19] and string theory [25] . These theories have a common denominator. It is assumed that the characteristics of the electron and atomic nucleus, such as charge, fields, forces and energy, are intrinsic properties. Hence, today's physics claims that they are generated without external influence [12] [20] - [24] . The exception is the Higgs Field and the Higgs Boson [9] - [11] .

This paper proposes that it is quite the opposite. The atom in itself contains no forces or energy, what so ever. The atom is completely passive. All energy in the universe originates from a singularity, positioned at the centre of the universe. This energy is absorbed by the electron and the atomic nucleus and the absorbed and re-emitted energy result in phenomena observed in physics: stored energy, forces, current and fields.

This section is structured in the following way. It begins with a model describing the electron's processes which create charge, Coulomb force, current, electric field and magnetic field. Only two quanta are involved: the energy quantum and the oscillator quantum. Both have a frequency specific to the electron.

The same model is then utilized to explain the nucleus strong force. The strong force is created by a process similar to the Coulomb force, albeit by energy quanta and oscillator quanta with frequencies specific to the nucleus and the strong force.

The same model is once again utilized to explain gravity. Gravity is created by a process similar to the Coulomb force, albeit by energy quanta and oscillator quanta with frequencies specific to gravity.

The same model is finally used to explain photons and inertial mass. Photons and inertial mass are created by a process where enforced energy hv is superpositioned on energy quanta and where the mechanism enabling this conversion resides in oscillator quanta.

Singularity

According to the theoretical model the universe contains a singularity. This singularity has been described previously [1] [2] [4] - [8] . This singularity generates energy quanta at six different frequencies. The energy quantum has no mass and propagates with the speed of light. An energy quantum is described by its vector δS_v in the present study, where v denotes its frequency v. The energy quantum δS_v remains unaltered as it travels through space. The singularity emits energy quanta δS_v uniformly into space creating energy flow density $S_v(0)$ at the singularity and energy flow density $S_v(R)$ at the distance R. Energy quanta δS_v are radiated uniformly into space, which results in that $S_v(R)$ decreases with the area of the sphere

$$S_v(R) = S_v(R = 0)\frac{1}{4\pi R^2}.$$

(1)

Electron Charge

This section describes the process which results in electron charge and Coulomb force. The process is general and is, in Sections 2.7, 2.8 and 2.9, utilized to explain positron, nucleus and gravitational forces. The electron contains an (forced damped) oscillator with natural frequency v_e which absorbs energy quanta δS_{v_e} and energy flow density $S_{v_e}(R)$ with frequency v_e, originating from the singularity. e denotes electron. The oscillator can be described as an oscillator quantum. The absorbed energy quanta δS_{v_e} are re-emitted and distributed uniformly into space as energy flow density $S_{v_e}(r)$ at the distance r from the electron. Hence, the energy flow density emitted by the electron is

$$S_{v_e}(r) = \alpha S_{v_e}(R = 0)\frac{1}{4\pi R^2} \cdot \frac{1}{4\pi r^2},$$

(2)

where α is a constant which mirrors the oscillator quantum transfer function. However, at the earth the distance R to the singularity is constant. Equation (2) can, thus, be simplified into

$$S_{v_e}(r) = S_{v_e}(r=0)\frac{1}{4\pi r^2}.$$
$$(3)$$

From now on $S_{v_e}(r=0)$ is denoted $S_{v_e}(0)$. Equation (3) describes the energy flow density from one electron. Now one electron is positioned at the distance r from a second electron. There is a mutual flow of energy quanta δS_{v_e} and energy flow density $S_{v_e}(r)$ between the two electron's oscillator quantum, and where the energy flow density is described by Equation (3). Energy quanta δS_{v_e} propagating in opposite directions between the two electrons results in the repelling force δF_{v_e} on each electron's oscillator quantum

$$\delta F_{v_e}(r) = \delta S_{v_e} \cdot \delta S_{v_e}.$$
$$(4)$$

Energy flow density $S_{v_e}(r)$ propagating in opposite directions between an electron positioned at the distance r and an electron positioned at r = 0 results in a repelling force $F_{v_e}(r)$ on each electron's oscillator quantum

$$F_{v_e}(r) = S_{v_e}(r) \cdot S_{v_e}(0),$$
$$(5)$$

and which according to Equation (3) can be re-written into

$$F_{v_e}(r) = S_{v_e}(0) \cdot S_{v_e}(0)\frac{r}{4\pi r^3}.$$
$$(6)$$

Equation (6) can be compared with classical electromagnetic theory and Coulomb's law for two electrons with elementary charge e [12]

$$F_{Coulomb}(r) = \frac{e \cdot e}{4\pi \varepsilon_0}\frac{r}{r^3},$$
$$(7)$$

and where we see that $S_{v_e}(0)$ corresponds to the classical definition of the electron elementary charge e, 1.6019×10^{-19} coulombs. ε_0 denotes that Equation (7) has been calculated using the electric field E instead of displacement D, $D=\varepsilon_0 E$ (see Section 2.5). Equation (6) can be modified to include clusters of k electrons with energy flow density $\sum_1^k S_{v_e}(r)$ or $\iint S_{v_e}(r)$ and which is proportional to the number of oscillator quanta involved. $\sum_1^k S_{v_e}(0)$ corresponds to the charge Q_1 or Q_2. This corresponds to Coulomb's law for two charges at distance r [12]

$$F_{Coulomb}(r) = \frac{Q_1 Q_2}{4\pi \varepsilon_0}\frac{r}{r^3}.$$
$$(8)$$

Electron Current

Current i_{v_e} is the transport with velocity v of the re-emitted energy flow density $S_{v_e}(0)$ from n electrons:

$$i_{v_e} = -n \cdot S_{v_e}(0) \cdot v,$$

(9)

where the "?" sign denotes that the definition of current flow is opposite to the electron transport direction. Equation (9) can be compared with the classical definition of current [12]

$$i = -n \cdot e \cdot v,$$

(10)

where $S_{v_e}(0)$ corresponds to the elementary charge e as described in Section 2.3.

Electron Magnetic Field

In the present study magnetism is described in the following way. An electron, with velocity v, passes a fixed point (x, y, z) in a given reference system. The electron re-emits the energy flow density $S_{v_e}(r)$. When the electron passes the fixed point (x, y, z) the derivate of $S_{v_e}(r)$ is observed, i.e. $\frac{\partial S_{v_e}(r)}{\partial t}$. Perpendicular to the direction of v, the derivate $\frac{\partial S_{v_e}(r)}{\partial t}$ forms equipotential surfaces, called $H_{v_e}(r)$ at the distances r. Each equipotential surface is continuous; hence it is described by its curl

$$\mathrm{curl}\, H_{v_e}(r) = \frac{\partial S_{v_e}(r)}{\partial t}.$$

(11)

Equation (11) describes $H_{v_e}(r)$ resulting from one electron. $H_{v_e}^n(r)$ from n electrons propagating with the velocity v is then described by combining Equation (9) and Equation (11)

$$\mathrm{curl}\, H_{v_e}^n(r) = i_{v_e}.$$

(12)

Now we picture current i_{v_e} propagating in a wire. The wire is cut and at this place a capacitor is positioned. Electrons accumulate on each capacitor plate and each electron on one plate results in energy flow density $S_{v_e}(r)$ according to Equation (3). This energy flow density and its derivate also form equipotential surfaces. n electrons result in energy flow density $S_{v_e}^n(r)$ which is the vector sum of the energy flow density from n contributing electrons at the distance r from the capacitor plate. Hence, $H_{v_e}^n(r)$ at the distance r from the capacitor plate is

$$\text{curl}H_{v_e}^n(r) = \frac{\partial S_{v_e}^n(r)}{\partial t}.$$

(13)

Consequently, $H_{v_e}^n(r)$ resulting from current i_{v_e} propagating in the wire and energy flow density $S_{v_e}^n(r)$ propagating in vacuum (in the capacitor) is

$$\text{curl } H_{v_e}^n(r) = i_{v_e} + \frac{\partial S_{v_e}^n(r)}{\partial t}.$$

(14)

$H_{v_e}^n(r)$ can be defined as the magnetic field, similar to the magnetic field H in classical electromagnetic theory. Equation (14) corresponds to Maxwell's equation [12]

$$\text{curl}H = i + \frac{\partial D}{\partial t}.$$

(15)

$S_{v_e}^n(r)$ corresponds to the displacement D in classical electromagnetism and $S_{v_e}(r)$ corresponds to the displacement D from one electron (i.e. from one oscillator quantum). According to classical electromagnetism $D = \varepsilon_0 E$ (in vacuum). Thus, the electric field E corresponds to $\frac{S_{v_e}^n(r)}{\varepsilon_0}$. However, D and E are defined as static fields in classical electromagnetism. In the present theory they consist of energy quanta with frequency v_e.

Revised Electromagnetic Theory

It is proposed that the electromagnetic theory can be revised and simplified. The revised version includes energy quantum δS_{v_e}, oscillator quantum with frequency v_e, energy flow density from one electron $S_{v_e}(r)$, energy flow density from n electrons $S_{v_e}^n(r)$ (displacement D), force $F_{v_e}(r)$ according to Equation (5), current i_{v_e} according to Equation (9), magnetic field $H_{v_e}^n(r)$ according to Equation (14) and voltage V $(E = -\text{grad}V)$. The electric field E can be replaced with $D = \varepsilon\varepsilon_0 E$. The magnetic field $B_{v_e}^n(r)$ can be replaced with $B_{v_e}^n(r) = \mu\mu_0 H_{v_e}^n(r)$. Maxwell's equations are still valid, provided that they are modified using energy flow density and its time derivate according to the above description.

Sections 2.7, 2.8 and 2.9 explain how the revised electromagnetic theory can be utilized to describe the positron, the nucleus, gravity and astrophysics.

The revised electromagnetic theory is general and forms the fundament for unification of many scientific disciplines.

Positron

This section explains that the positron is identical to the electron; however, the positron energy quantum and oscillator quantum have a frequency specific to the positron. The positron contains an (forced damped) oscillator quantum with natural frequency v_p which absorbs energy quanta δS_{v_p} and energy flow density $S_{v_p}(R)$, and where p denotes positron. Thus, the positron frequency is different to the electron frequency. The absorbed energy consists of energy quanta δS_{v_p} which are re-emitted and distributed uniformly into space as energy flow density $S_{v_p}(r)$ at the distance r from the positron. The rules and laws of the positron are identical to those of the electron, except that its frequency is v_p. There is a repelling force caused by energy flow density $S_{v_p}(r)$ between two positrons

$$F_{p_p}(r) = S_{v_p}(0) \cdot S_{v_p}(0) \frac{r}{4\pi r^3}.$$

(16)

Hence, energy flow density between oscillators having exactly the same frequency results in repelling force. However, energy flow density between oscillators having slightly different frequencies results in attracting force.

The electron and the positron have slightly different frequencies. Thus, there is a force of attraction $F_{e_p}(r)$ between energy flow density $S_{v_e}(r)$ from an electron operating on the positron oscillator quantum with frequency v_p. Likewise, there is a force of attraction $F_{e_p}(r)$ between energy flow density $S_{v_p}(r)$ from a positron operating on the electron oscillator quantum with frequency v_e.

$$F_{e_p}(r) = -S_{v_e}(0) \cdot S_{v_p}(0) \frac{r}{4\pi r^3},$$

(17)

where the "−" sign denotes force of attraction. Consequently, the positron applies to revised electromagnetic theory and revised Maxwell's equations provided parameters include the positron frequency v_p.

Strong Force and Nuclear Energy

This section explains that the atomic nucleus contains mechanisms similar to the positron and electron; however, the nucleus energy quanta and oscillator quanta have frequencies specific to the nucleus. Neutrons and protons contain large numbers of oscillator quanta with two slightly different frequencies

v_{n1} and v_{n2}, where n denotes nucleus and its strong force. This creates energy flow density between oscillator quanta as described in Sections 2.3 and 2.7. The energy flow density results in force of attraction $F_{v_{n12}}(r)$ between oscillator quanta with slightly different frequencies, v_{n1} and v_{n2}, within fermions, within neutrons, within protons and between neutrons and protons.

$$F_{v_{n12}}(r) = -S_{v_{n1}}(r) \cdot S_{v_{n2}}(0). \tag{18}$$

It also creates repelling forces $F_{v_{n11}}(r)$ and $F_{v_{n22}}(r)$ between oscillator quanta with equal frequency v_{n1} or v_{n2}

$$F_{v_{n11}}(r) = S_{v_{n1}}(r) \cdot S_{v_{n1}}(0). \tag{19}$$

It is proposed that $F_{v_{n12}}(r)$ is larger than $F_{v_{n11}}(r)$ and $F_{v_{n22}}(r)$, creating a net force of attraction. This implies that the strong force between oscillators in the nucleus is similar to the force between electrons and positrons, except the difference in frequencies. It is proposed that oscillators in the nucleus are not static; they oscillate and orbit each other similar to electrons orbiting the nucleus. Fermions, neutrons and protons contain very large numbers of oscillator quanta; consequently, the combined forces result in very strong forces. The oscillators are densely packed, resulting in that re-emitted energy density $S_{v_{n1}}(r)$ and $S_{v_{n2}}(r)$ are primarily absorbed by oscillators within the nucleus and minor energy flow density outside the nucleus. This also implies that the energy flow density within the nucleus is very large. This results in stored energy, i.e. nuclear energy and binding energy within fermions, within neutrons, within protons and between neutrons and protons.

It is proposed that elementary particles like fermions, neutrons and protons are different configurations of oscillators with frequency $v_{n1}, v_{n2}, v_e, v_p,$ v_{g1} and v_{g2}, where v_{g1} and v_{g2} are described in Section 2.9. Bosons are energy quanta with frequency $v_{n1}, v_{n2}, v_e, v_p, v_{g1}$ and v_{g2}.

The implication of this is that the atomic nucleus can and should be described by the same laws as the revised electromagnetic theory, except that the frequencies are different (v_{n1} and v_{n2}). Consequently, the nucleus is described by the nucleus energy quantum vectors δS_{v_n}, nucleus energy flow density from one oscillator $S_{v_n}(r)$, nucleus energy flow density from n oscillators $S_{v_n}^n(r)$ (nucleus displacement D_{v_n}), nucleus force $F_{v_n}(r)$ according to Equation (5), nucleus current i_{v_n} according to Equation (9), and nucleus magnetic field $H_{v_n}^n(r)$ according to Equation (14). The nucleus is not static; oscillators in fermions, neutrons and protons orbit and oscillate. This creates internal displacement D_{v_n} and magnetic field $H_{v_n}^n(r)$ which may result in what is called the nuclear weak force. The atomic nucleus applies to the revised electromagnetic theory, provided parameters include the nucleus frequencies $v_{n1}, v_{n2}, v_e, v_p, v_{g1}$ and v_{g2}.

Gravity

This section explains that gravity is created by mechanisms similar to the electron, positron and the atomic nucleus; however, gravity encompasses energy quanta and oscillator quanta with frequencies specific to gravity. Gravity has been described in detail in [2] . Neutrons, protons and electrons contain oscillator quanta with natural frequencies v_{g1} and v_{g2}, where g denotes gravity. The frequencies v_{g1} and v_{g2} are slightly different. This creates energy flow density between oscillator quanta as described in Sections 2.3 and 2.7. The energy flow density creates force of attraction $\delta F_{v_{g12}}(r)$ between oscillator quanta in atoms and electrons having slightly different frequencies v_{g1} and v_{g2}

$$\delta F_{v_{g12}}(r) = -S_{v_{g1}}(r) \cdot S_{v_{g2}}(0).$$

(20)

The force of attraction $F_{v_{g12}}(r)$ between two bodies containing n respectively p oscillator quanta is

$$F_{v_{g12}}(r) = -nS_{v_{g1}}(r) \cdot pS_{v_{g2}}(0).$$

(21)

It is assumed that that the distance r between the bodies is much larger than the distance between oscillator quanta within each body. Equation (21) can be re-written into

$$F_{v_{g12}}(r) = -\gamma np \cdot S_{v_{g1}}(0) \cdot S_{v_{g2}}(0)\frac{r}{4\pi r^3} = -\eta np \cdot \frac{r}{4\pi r^3},$$

(22)

where γ and η are constants. It also creates repelling forces $F_{v_{g11}}(r)$ and $F_{v_{g22}}(r)$ between oscillator quanta with equal frequency v_{g1} or v_{g2}

$$F_{v_{g11}}(r) = \xi\gamma np \cdot S_{v_{g1}}(0) \cdot S_{v_{g1}}(0)\frac{r}{4\pi r^3} = \xi\eta np \cdot \frac{r}{4\pi r^3}.$$

(23)

ξ denotes that the repelling forces $F_{v_{g11}}(r)$ and $F_{v_{g22}}(r)$ are smaller than $F_{v_{g12}}(r)$. This creates a net force of attraction $F_g(r)$. ξ can be described as $1 - \Delta$, where Δ is extremely small. This explains why the gravitational force $F_g(r)$ is extremely small. The number of oscillator quanta n in body 1 respectively p in body 2 is proportional to the gravitational mass m_1 of body 1 and the gravitational mass

m_2 of body 2. We now recognize the similarity with Newton's metric law of gravity, where G is Newton's gravitational constant [13].

$$F_g(r) = \Delta\eta np \cdot \frac{1}{4\pi r^2} \approx F_{\text{Newton}} = Gm_1 m_2 \frac{1}{r^2}.$$

(24)

We also recognize the similarity with Coulomb's law $F_{\text{Coulomb}}(r) = \frac{Q_1 Q_2}{4\pi\varepsilon_0 r^2}$. The reason is that the gravita-

tional force is similar to the force between clusters of electrons. The number of involved oscillator quanta in atoms and the frequencies v_{g1} and v_{g2} determines the gravitational constant $\Delta\eta$.

The implication of this is that the solar system, the Milky Way Galaxy and the universe in general should be described by the same laws as the revised electromagnetic theory, except that the frequencies are different (v_{g1} and v_{g2}). Hence, the relationship between planets, stars and galaxies is described by the gravitational energy quantum vector δS_{v_g}, gravitational energy flow density from one oscillator quantum $S_{v_g}(r)$, gravitational energy flow density from n oscillator quanta $S_{v_g}^n(r)$ (gravitational displacement D_{v_g}), gravitational force $F_{v_g}(r)$ according to Equation (5), gravitational current i_{v_g} according to Equation (13), and gravitational magnetic field $H_{v_g}^n(r)$ according to Equation (14). Hence, all bodies (planets, stars) in the universe generate gravitational current, gravitational force, gravitational displacement and gravitational magnetic field. They apply to the revised electromagnetic theory, provided the parameters include the gravitational frequencies v_{g1} and v_{g2}.

Photon

The nature of the photon has been described in detail [4]. The process creating photons is described here in condensed form. Particles, e.g. electron, neutron and proton contain one or many oscillator quanta with natural frequency v_g creating gravity as described in Section 2.9. Each oscillator quantum absorbs energy quanta δS_{vg} and re-emits the energy quanta δS_{v_g}. Accelerating an electron implies that force and energy is enforced on the electron. This energy can be described as many energy quanta hv. The energy quanta hv cannot be stored in the electron and its oscillator quanta. The energy quanta hv are, as a result, superpositioned on re-emitted energy quanta δS_{v_g}. One energy quantum hv is superpositioned (in the oscillator quantum) on one re-emitted energy quantum δS_{v_g} at a time. This new energy quantum δS_{v_g+v} has the frequency v_g and v, where v_g can be either v_{g1} or v_{g2}. Likewise, decelerating an electron implies that force and energy is enforced on the electron. The energy quanta hv are, therefore, superpositioned on re-emitted energy quanta in exactly the same way as when electrons are accelerated. Radio waves, light and gamma rays are energy quanta hv superpositioned on gravity energy quanta $^{\delta S_{v_g}}$.

Gravitational and Inertial Mass

Gravitational and inertial mass have been described in detail [5] . The process behind gravitational and inertial mass is described here in condensed form.

Gravitational mass is the expression of the mechanism, i.e. atomic oscillator quanta which absorb and re-emit energy quanta $^{\delta S_{v_g}}$ and energy flow density $S_{v_g}(r)$. This results in gravitational force according to Equations (22) and (23). Consequently, the gravitational mass m_1 in Equation (24) is proportional to the number n of oscillator quanta with natural frequency v_g (creating gravity) in the mass or the body.

Accelerating or decelerating a mass, a particle, an electron or an atom demands enforced force and energy. The mass, particle, electron or atom cannot store this enforced energy. The enforced energy can be described as energy quanta hv. The enforced energy quanta hv are superpositioned on absorbed and re-emitted energy quanta $^{\delta S_{v_g}}$ as described in Section 2.10. Hence, the enforced energy quanta hv are emitted as photons with energy hv. Thus, inertial mass is the expression of the mechanism, i.e. atomic oscillator quanta with natural frequency v_g, which convert enforced energy quanta hv into emitted photons with energy hv. The inertial mass is proportional to the number n of oscillator quanta (creating gravity) in the mass. Consequently, the inertial mass is proportional to the gravitational mass, although gravitational and inertial mass originate from totally different processes. Hence, the process creating inertial mass is identical to the process creating photons, as described in Section 2.10.

Quantum Physics

In this section it is assumed that the electron comprises three oscillator quanta, each one with frequencies v_e, v_{g1} and v_{g2} (see Section 3). This simple configuration of three oscillator quanta provides the electron with intrinsic characteristics described in Sections 2.3, 2.4, 2.5, 2.9, 2.10 and 2.11. The oscillator quanta v_e, v_{g1} and v_{g2} do not comprise a static configuration. They oscillate and orbit each other.

The electron has intrinsic angular momentum (up spin and down spin) and intrinsic magnetic dipole momentum, according to quantum physics [20] [21] . The oscillator quanta v_e, v_{g1} and v_{g2} orbit each other, i.e. they have spin. The energy quanta δS_{v_e}, $\delta S_{v_{g1}}$ and $\delta S_{v_{g2}}$ are polarized [1] [2] [4] . This polarization results in that the oscillator quanta spin is oriented either 0 degree or 180 degrees relative to the absorbed energy quanta. This is called up spin and down spin. The spin implies that all three oscillator quanta accelerate in their orbits. It results in both magnetic fields H_{v_e}, $H_{v_{g1}}$ and $H_{v_{g2}}$, as described in Section 2.5 and magnetic dipole momentums.

The electron has a wave function and applies to the wave-particle duality [20] [21] . Sections 2.3 and 2.9 describe that the electron absorbs and re-emits energy quanta δS_{v_e}, $\delta S_{v_{g1}}$ and $\delta S_{v_{g2}}$. The re-emitted energy flow density S_{v_e} is described in Equation (3). This defines the electrons characteristics. Consequently, the electron and its oscillator quanta is primarily a wave function comprising the frequencies v_e, v_{g1} and v_{g2}. Thus, the electron can be described as a particle consisting of its oscillator quanta as well as a wave consisting of re-emitted energy quanta. The electron's wave function explains why electrons orbiting the nucleus exhibit discrete orbits. The electron's permitted orbits, with radius r, are related to the electron's wavelength $\lambda_e = n2\pi r$, where $\lambda_e = 1/v_e$ and $n = 1,2,3,\text{L}$.

The electron is exposed to forces between the electron and the singularity as well as forces between the electron and adjacent electrons and protons. These forces are caused by absorbed energy quanta with frequencies v_e, v_{g1} and v_{g2}, as described in Equations (6), (17), (22) and (23). These forces are not static and, consequently, the electron's position will oscillate with the frequencies v_e, v_{g1} and v_{g2}. This may shed light on the uncertainty principle [20] [21] .

According to classical and quantum physics an orbiting (accelerated) electron radiates energy (photons) [20] [21] . Hence, it loses energy and should, consequently, rapidly approach and collide with the nucleus. However, electrons have fixed orbits. The electron absorbs energy quanta from the singularity and which compensates lost energy when the electron is accelerated in its orbit. Hence, there is energy balance which ensures fixed orbits.

The Pauli Exclusion Principle says that two fermions (e.g. two electrons) cannot occupy the same quantum state [20] [21] . This is illustrated with electrons orbiting the nucleus. In order to understand the Pauli Exclusion Principle injection locking is introduced [22] [23] . Injection locking is the frequency effects that can occur when an oscillator is disturbed (or frequency shifted) by injected energy. This implies that the electron oscillator quantum frequency v_e is slightly altered by energy (photons) generated when the electron is accelerated in its orbit (see Section 2.10). Altering the frequency v_e results in that the force of attraction $F_{e_p}(r)$, Equation (17), between electron and nucleus is slightly reduced, whereby the electron is forced into an orbit with a different radius. This orbit is stable when there is equilibrium between acceleration, v_e, $F_{e_p}(r)$ and when $\lambda_e = n2\pi r$ is fulfilled. Hence, the electron can have many stable and discrete orbits. In principle the radius can vary with time, resulting in diffuse; however, stable orbits. The repelling force F_{v_e}, Equation (6), between orbiting electrons is minimized when they have different states, i.e. different orbits. Hence, each orbit can be occupied by only one electron with the same quantum state, i.e. one electron with up spin and one with down spin.

Consequently, phenomena observed in quantum physics can be explained with the present model.

ELEMENTARY PARTICLES

In this section it is assumed that the universe is organized in a simple and logical way. It is proposed that the universe, outside the singularity, contains only two types of primary entities. The two primary entities are the energy quantum δS_v and the atomic/particle oscillator quantum with natural frequency v. They are the two universal quanta. They can have six different frequencies; v_e, v_p, v_{n1}, v_{n2}, v_{g1} and v_{g2}. Each frequency provides unique characteristics to the electron, positron, nucleus, gravity and energy transport in space.

The six frequencies are arranged in the frequency spectrum in the following order: v_{n1}, v_{n2}, v_{g1}, v_{g2}, v_p, v_e, where v_{n1} has the highest frequency. These frequencies are very close to each other which results in that there is a force of attraction between oscillator quanta v_{n1}, v_{n2}, v_{g1} and v_{g2} as described in Equation (17). It is proposed that these forces tend to unite oscillator quanta into stable triplets of oscillator quanta, where one triplet consists of v_{n1}, v_{g1} and v_{g2} and the other triplet consists of v_{n2}, v_{g1} and v_{g2}. The frequencies v_{g1}, v_{g2}, v_p, v_e are also close to each other. It is proposed that it results in one triplet v_p, v_{g1} and v_{g2} (positron) and one triplet v_e, v_{g1} and v_{g2}(electron). There could, of course, be other configurations, e.g. v_{n1}, v_{n1}, v_{g1} and v_{g2}. However, the frequencies v_{n1} and v_e are too separated to form a bond according Equation (17). Thus, only some configurations are feasible. In the following a very simple structure is proposed. The purpose is to show that the atom may have simple and repetitive structures. It is not intended as a solution; however, a guideline showing possible principles.

In its simplest configuration the universe may consists of only four basic particles: electron (v_e, v_{g1}, v_{g2}), positron (v_p, v_{g1}, v_{g2}) and two nuclear basic particles called α-tron (v_{n1}, v_{g1}, v_{g2}) and β-tron (v_{n2}, v_{g1}. v_{g2}). The α-tron and β-tron are the enablers of the strong force and nuclear energy. All four basic particles have almost identical structures. They contain one common part (v_{g1}, v_{g2}) resulting in gravity, gravitational mass, inertial mass and production of photons. They also contain a second individual part consisting of one of the oscillator quanta v_{n1}, v_{n2}, v_p or v_e. This part results in the "positive" positron (v_p), the "negative" electron (v_e), the "positive" α-tron (v_{n1}) and the "negative" β-tron (v_{n2}). Thus, the four basic particles are identical except for one parameter; the frequencies v_{n1}, v_{n2}, v_p and v_e. The frequency parameter determines the characteristics of the different parts of the atom. Configurations of α-trons, β-trons, positron and electron constitute neutron, proton and other particles.

Perceived particle mass is always the result of oscillator quanta v_{g1}, v_{g2} and their re-emitted energy quanta δS_{v_g}. All forces are created by the oscillator quanta v_e, v_p, v_{n1}, v_{n2}, v_{g1}, v_{g2} and their re-emitted energy quanta. All nuclear energy is created by the oscillator quanta v_e, v_p, v_{n1}, v_{n2}, v_{g1}, v_{g2} and absorbed and re-emitted energy quanta. Fermions contain oscillator quanta. Bosons are energy quanta δS_v emitted by the singularity, absorbed by oscillator quanta in particles or re-emitted by oscillator quanta in particles.

The present model separates between the entity which enables fields, forces, stored energy and emitted energy (photons) on one hand and the entity which mediates fields, forces, stored energy and emitted energy on the other hand. The enablers of fields, forces, stored energy and emitted energy are the oscillator quanta v_e, v_p, v_{n1}, v_{n2}, v_{g1} and v_{g2}. The mediators of fields, forces, stored energy and emitted energy are the energy quanta δS_v with frequencies v_e, v_p, v_{n1}, v_{n2}, v_{g1} and v_{g2}.

DISCUSSION AND CONCLUSIONS

The present study proposes that all energy in the universe originates from a singularity at the centre of the universe. The universe, outside the singularity, contains only two types of primary entities. They are the energy quantum δS_v and the atomic/particle oscillator quantum with natural frequency v. They can have six different frequencies; v_e, v_p, v_{n1}, v_{n2}, v_{g1} and v_{g2}. Configurations of oscillator quanta with different natural frequencies constitute particles such as fermions, electron, positron, neutron and proton. Each frequency provides unique characteristics to electron, positron, nucleus, gravity and energy transport in space. Energy transport in space is achieved by energy quanta δS_v or by superpositioned energy quanta hv (photons) on energy quanta δS_{v_g}.

According to this study electromagnetic theory, quantum physics, elementary particle physics and astrophysics can be described by the same set of rules and laws. The revised version of electromagnetic theory constitutes the fundament for unification of these scientific disciplines into one comprehensive Theory of Everything (TOE). The novel Theory of Everything will dramatically simplify the description of many natural sciences and ease the understanding of these sciences. Today there are often locked doors between electromagnetic theory, quantum physics, elementary particle physics, astrophysics, bioelectromagnetics, chemistry, biochemistry and medicine. The reason is, among others, that these discipline use different models and different terminology. The Theory of Everything will unify models and terminology. It will result in increased multidisciplinary cooperation and faster progress in R&D.

The electronics industry rapidly approaches processes and products which operate on only one or few electrons. Understanding the real nature of the electron and the photon can result in a paradigm shift within microelectronics.

Bioelectromagnetics is an important discipline, concerned with the nature of organisms and which has large impact on medicine. The Theory of Everything will dramatically change the definition of bioelectromagnetics processes from today's static processes to dynamic electromagnetic processes. It will improve the understanding of the human organism and infectious diseases. It may result in a paradigm shift in medicine.

Many environmental problems are caused by chemicals used in e.g. industrial processes, products and in farming, and where the chemicals eventually are found in the environment, in food and in the human organism. They consist of molecules called PFOS, PFOA, PCB, PBDE, glyphosate, carbendazim, boscalid etc. Molecules are today described as atoms united by the static Coulomb force. In reality the molecule shall be described by the Theory of Everything; energy quanta and oscillator quanta in the atomic nucleus and the electron result in a number of forces which amalgamate atoms into a molecule. This implies that the molecule comprises many dynamic electromagnetic processes. These dynamic electromagnetic processes interact with and disturb dynamic electromagnetic processes in organisms including humans, i.e. the processes described in bioelectromagnetics. Today's static definition of the molecule provides meager understanding of the environmental effects caused by chemicals. In the future it may be possible to disturb the molecule's dynamic processes; it results in that the molecule is destroyed. It may enable remediation of the environment from toxic chemicals. The Theory of Everything may cause a paradigm shift in chemistry and in the understanding of environmental problems.

The earth is exposed to an enormous amount of energy, i.e. energy quanta δS_v. Only a fraction of these energy quanta are absorbed by atoms. The absorbed energy results in nuclear fission and fusion energy used in e.g. nuclear power plants. It may be possible to control this vast amount of energy in the future.

There may be no energy problem in the future; there will be endless amount of energy and the energy will be safe with no waste and without environmental impact.

It is proposed that the present study will have major impact on the understanding of basic physics, chemistry and medicine. Hopefully, this will result in multidisciplinary cooperation, faster R&D, improved products, improved medicine, improved understanding of environmental problems, improved methods to remediate the environment and a solution to the energy and climate problem.

REFERENCES

1. Giertz, H.W. (2013) Open Journal of Microphysics, 3, 115-120.

2. Giertz, H.W. (2013) International Journal of Astronomy and Astrophysics, 3, 39-50.http://dx.doi.org/10.4236/ijaa.2013.32A007

3. Giertz, H.W. (2010) Journal of Atmospheric and Solar-Terrestrial Physics, 72, 767-773.http://dx.doi.org/10.1016/j.jastp.2010.03.022

4. Giertz, H.W. (2013) Open Journal of Microphysics, 3, 71-80.http://dx.doi.org/10.4236/ojm.2013.33013

5. Giertz, H.W. (2014) Open Journal of Microphysics, 4, 7-14.http://dx.doi.org/10.4236/ojm.2014.42002

6. Giertz, H.W. (2014) International Journal of Astronomy and Astrophysics, 4, 353-358.http://dx.doi.org/10.4236/ijaa.2014.42029

7. Giertz, H.W. (2014) International Journal of Astronomy and Astrophysics, 4, 294-300.http://dx.doi.org/10.4236/ijaa.2014.41024

8. Giertz, H.W. (2014) International Journal of Astronomy and Astrophysics, 4, 384-390.http://dx.doi.org/10.4236/ijaa.2014.42033

9. Higgs, P.W. (1964) Physical Review Letters, 11, 508-509.http://dx.doi.org/10.1103/PhysRevLett.13.508

10. Higgs, P.W. (1964) Physical Letters, 12, 132-133. http://dx.doi.org/10.1016/0031-9163(64)91136-9

11. Englert, F. and Brout, R. (1964) Physical Review Letters, 13, 321-323. http://dx.doi.org/10.1103/PhysRevLett.13.321

12. Bleaney, B.I. and Bleaney, B. (1965) Electricity and Magnetism. Oxford University Press, Amen House, London.

13. Weinberg, S. (1972) Gravitation and Cosmology. John Wiley & Sons, Hoboken.

14. Okun, L.B. (1989) Physics Today, 42, 31-36. http://dx.doi.

org/10.1063/1.881171

15. Kneubühl, F.K. (1997) Oscillations and Waves. Springer, Berlin.http://dx.doi.org/10.1007/978-3-662-03468-2

16. Sipe, J.E. (1995) Physical Review A, 52, 1875-1883.http://dx.doi.org/10.1103/PhysRevA.52.1875

17. Oda, I. (2010) Physics Letters B, 690, 322-327.http://dx.doi.org/10.1016/j.physletb.2010.05.048

18. Bezrukov, F. and Shaposhnikov, M. (2008) Physics Letters B, 659, 703-706.http://dx.doi.org/10.1016/j.physletb.2007.11.072

19. Einstein, A. (1916) Annalen der Physik, 49, 285-339.

20. Eisberg, R. and Resnick, R. (1985) Quantum Physics of Atoms, Molecules, Solids, Nuclei, and Particles. 2nd Edition, Wiley, Hoboken.

21. Harris, R. (2008) Modern Physics. 2nd Edition, Pearson International Edition, Addison-Wesley, Boston.

22. Kurokawa, K. (1973) Proceedings of the IEEE, 61, 1386-1410.http://dx.doi.org/10.1109/PROC.1973.9293

23. Tiebout, M. (2004) IEEE Journal of Solid-State Circuits, 39, 1170-1174. http://dx.doi.org/10.1109/JSSC.2004.829937

24. Burgess, C. and Moore, G. (2007) The Standard Model: A Primer. Cambridge University Press, Cambridge.

25. Polchinski, J. (1998) String Theory. Cambridge University Press, Cambridge.

Chapter 4

SYMMETRY AND RELATIVITY: FROM CLASSICAL MECHANICS TO MODERN PARTICLE PHYSICS

Z. J. Ajaltouni

Laboratoire de Physique Corpusculaire de Clermont-Ferrand IN2P3/CNRS Université Blaise Pascal F-63177, Aubière Cedex, France;

ABSTRACT

The aim of this review is to highlighte the common aspects between Symmetry in Physics and the Relativity Theory, particularly Special Relativity. After a brief historical introduction, emphasis is put on the physical foundations of Relativity Theory and its essential role in the clarification of many issues related to fundamental symmetries. Their different connections will be shown from Classical Mechanics to Modern Particle Physics.

INTRODUCTION

Physical sciences rely essentially on two pillars: experimental facts, on one side, and their translation into a coherent mathematical formalism on another side. In both the two approaches, a physical insight is used, symmetry. Does the physical process under analysis present symmetrical properties, or does its formulation contain some symmetry elements? This kind of intellectual attitude has been adopted in Physical Sciences since the famous paper of P. Curie written at the end of the 19th century in which the author stresses on symmetry properties of some ectromagnetic phenomena, notably those ones occurring in crystal bodies [1].

Quoting H. Weyl's sentence summarized by: "Symmetry is one idea by which man through the ages has tried to comprehend and create order, beauty, and perfection..." ([2]), one can add that symmetry represents a methodology followed by Modern Physics in order to build coherent and successful models whose aim is to understand the fundamental physical laws at all scales, from the microscopic world to the macroscopic Universe.

Let us come to Relativity Theory and how it is related to Symmetry in Physics. Firstly, one has to remind that Relativity is born from the unsufficiencies of Classical Physics, both Newton Mechanics and Maxwell Electromagnetism, in which the absolute character of time and the existence of a hypothetic medium—the aeter, are postulated, which fills the vacuum and serves as a support for the propagation of electromagnetic waves, among them the visible light.

After the publication of A. Einstein's historical paper [3], time like space components becomes an ordinary component of a four-dimensional structure, the Minkowski space-time manifold. Thus, a complete symmetry arises between space and time, and this new feature leads to important physical consequences as it will be shown in the next sections.

This review is organized around three main topics:

- Contribution of Relativity Theory to the different symmetries in Classical Physics (Section 2).
- Relativistic Symmetries in Quantum Mechanics (Section 3).
- Role of the Relativity Theory in the discovery of new symmetries (and asymmetries) in Elementary Particle Physics (Section 4).

The conclusion will be devoted to some open issues in the common field of Relativity and Symmetry.

It is worth noticing that the different parts of the review are related to each other by two motivations: 1) the first one is the evolution of the physical concepts initiated by Relativity Theory, and 2) the second one is the breakthrough that some discoveries brought to the Physical Sciences, notably in our understanding of the Symmetry concept.

RELATIVITY AND SYMMETRIES IN CLASSICAL PHYSICS

Classical Mechanics

One of the main difficulties which appeared in Classical Mechanics was its inconsistency with the electromagnetic phenomena, notably the problem of magnetic induction in a moving circuit. This inconsistency has been amplified by the negative result of the Michelson Morley experiment whose aim was measuring the translation velocity of the Earth in the ether. In order to cure these problems, A. Einstein succeeded in establishing new basis for Mechanics by setting three new postulates:

- Time is a relative notion and each observer possesses its own time.
- The speed of light is isotropic and does not depend on the state of motion of the observer.
- (R') being an inertial frame moving with a constant velocity v according to another frame (R), the coordinate transformation between these two inertial frames is given by the Lorentz formula [4].

A straightforward consequence of these new principles was the fusion of space and time in one entity called "Space-Time", which was performed by H. Minkowski ([5]). In this new 4-space, time and space components are completely symmetric, as it is shown by the following series of equalities relating space-time components of the same event between the two frames (R) and (R'):

$$x = x', y = y', z = \gamma(z' + vt'), t = \gamma\left(t' + vz'/c^2\right)$$

(1)

and

$$x' = x, y' = y, z' = \gamma(z - vt), t' = \gamma\left(t - vz/c^2\right)$$

(2)

where $\gamma = \dfrac{1}{\sqrt{1 - v^2/c^2}}$ is the Lorentz factor, and the sign $-$ appears in Equation (2) because the velocity of (R) according to (R') is $-v$.

Physical interpretation of these relations becomes more obvious by adopting the point of view of Special Relativity: Analysis of physical phenomena does not depend on the motion of the observer(s). Observer O in frame (R) and observer O' in frame (R') have completely a symmetric role. It is worth stressing that Einstein's postulates and Minkowski's 4-space lead to a new physical concept, the Relativistic Invariance one, which have important consequences in the foundations of Modern Physics.

Thanks to a new mathematical entity, the 4-vector introduced by Minkowski, one can set many fundamental invariants, among them:

- The square of the space-time interval, $ds^2 = dx''dx_\mu = c^2 dt^2 - dx^2 - dy^2 - dz^2$, from which one can deduce automatically the relation of clock retardation [6].

- The square of the energy-momentum 4-vector, $P^\mu P_\mu = E^2/c^2 - p^2 = m^2 c^2$.

Applying this last relation to a massive particle at rest $(p = 0)$, the Einstein breakthrough formula emerges, $E_{(0)} = mc^2$, which represents a new symmetry, the equivalence between Energy and Mass [7]; symmetry which does not exist in Newton Mechanics.

Classical Electrodynamics

In his seminal paper, A. Einstein was motivated by showing that the two standard views of the induction phenomenon (Lenz-Faraday law) were really the same one. Departing from the Relativity principles, he achieved merging the two different processes and he found again the standard results of the Maxwell's equations: any variable electric field E generates a magnetic field B, and reciprocally. So, a complete symmetry is established between the electric and magnetic fields, despite the absence (temporary?) of magnetic charges. This physical property is clearly reflected in the relativistic formalism with the introduction of a new entity, the electromagnetic tensor $F_{\mu\nu} = \partial_\mu A_\nu - \partial_\nu A_\mu$ [1], which is an antisymmetric tensor whose six components are respectively those of the electric E and magnetic B fields. Doing some calculations with this new tensor like the following, $\partial_\alpha F_{\beta\gamma} + \partial_\gamma F_{\alpha\beta} + \partial_\beta F_{\gamma\alpha} = 0$ (indices α, β, γ are varying from 0 to 3), allows to find again the standard Maxwell-Faraday equations: $\text{rot}E = -\partial B/\partial t$ and $\text{div}B = 0$.

Summary

At the classical level, the Relativity Theory introduces new aspects of symmetry in fields where symmetry was not conjectured: Space and Time, Electric and Magnetic fields and the Mass-Energy equivalence. Furthermore, its mathematical formulation includes, in a simple and direct way, the symmetry properties of the physical processes [8].

RELATIVITY AND THE SYMMETRY PROPERTIES OF QUANTUM MECHANICS

We will deal successively with non-relativistic Quantum Mechanics (NRQM), relativistic Wave Mechanics of De Broglie and the relativistic approach to Quantum Mechanics initiated by P. Dirac. By following the historical steps, emphasis is put on the basic physical ideas and the main experimental discoveries in each field. Technical calculations will be avoided.

Symmetry in Non-Relativistic Quantum Mechanics

The importance of Symmetry in NRQM leads to a wide use of Group Theory methods when describing quantum systems, from the simple hydrogen atom to complicated molecular structure [9]. This feature is related to the mathematical structure of Quantum Mechanics itself: any physical quantity which can be measured, or observable, is represented by a hermit an operator $A\left(A=A^{\dagger}\right)$ acting on a Hilbert space built from state vectors or kets; the last ones represent the physical states of the system and form an algebraic vector space [10]. How ordinary QM has been made relativistic? Before answering this question, it is fair to recall how Special Relativity has been firstly introduced in Quantum Physics.

De Broglie Relativistic Wave Mechanics

In his doctoral thesis (Paris, 1924), the french physicist L. de Broglie generalized the important notion of wave-particle duality (first introduced by A. Einstein for the Photon in 1905 [11]) by extending it to massive relativistic particles [12]. Departing from the hypothesis that each material corpuscle has an internal vibration whose frequency is given by the Planck-Einstein relation, $hv = \gamma mc^2$, and supposing that this vibration is in phase with a "phase wave" propagating at the speed $V = c^2/v$ (²), De Broglie inferred the expression of the particle wave-length:

$$\lambda = \frac{V}{v} = \frac{h}{\gamma m v} = \frac{h}{p}$$

which is identical to the photon wave-length given by the Planck-Einstein relation, $E = pc = hv \Rightarrow \lambda = c/v = h/p$. This famous relation, which has been confirmed by the experimental discovery of the electron diffraction (Davisson and Germer, 1927), links together the two main aspects of matter, wave and/or particle. Thus, a new symmetry at the level of the elementary particles has been revealed, as a consequence of the Einstein-De Broglie relations. Then, the wave nature of any quantum system has been exploited very deeply to set fundamental equations: the non-relativistic Schrodinger equation [13], and the relativistic Klein-Gordon one [14].

Dirac Equation and Its Consequences

In the years 1927-1928, the British physicist P. Dirac succeeded in formulating a relativistic wave equation for the electron characterized both by its simplicity and its ingenuity [15].

$$\gamma^{\mu} p_{\mu} \Psi = mc\Psi$$

where $\left(p_{\mu}\right)$ is the 4-momentum of the particle and γ^{μ} are the Dirac matrices with $\mu = 0,1,2,3$. The Dirac equation has a linear evolution according to time, like the Schrodinger equation and, furthermore, it reveals a perfect similarity among space and time components. Thus, it has the great advantage to be relativistic invariant. Applied to the known massive particles (at this epoch there were only electrons and protons), Dirac equation leads to fascinating results:

- The value of the spin, $s = 1/2$, is deduced automatically with a gyro-magnetic factor of the electron, $g_e = 2$, which agrees with the experimental measurements.

- There appear negative-energy states for the electron• $E = -\sqrt{p^2 + m^2}$, as with the Klein-Gordon equation. But, according to the fermionic nature of the electron, these states are interpreted as an electron with positive electric charge, or an anti-electron called also a positron [16]. This particle has been discovered by Anderson in 1933 [17], which marks the beginning of the Antimatter Era.

- Thanks to the relativistic Dirac equation a new symmetry is born: and it confirmed the prediction of Dirac that "every particle has its own antiparticle, whatever its spin is".

Matter $\emptyset$ Antimatter

Then, there were the discoveries of the antiproton, the antimuon, the antineutron and the other ones [18].

- What is really fundamental in this new kind of symmetry is that it firstly appeared in the world of elementary particles, a field governed by the rules of Quantum Mechanics, and not in the macroscopic world where Classical Physics dominates.

It is worth saying too that taking account of the importance of the Relativity principles in both the two approaches to Quantum Physics, De Broglie's Wave Mechanics or Standard (Dirac) Quantum Mechanics, have paved the way to the development of a new field of research, the physics of Elementary Particles.

MODERN PARTICLE PHYSICS

At the same period than the Dirac equation, Quantum Field Theory (QFT) was developed on relativistic principles [19]. It was essentially devoted to the interactions between quantized electromagnetic fields and charged particles, which is called Quantum Electrodynamics (QED). The formal language of QFT uses intensively Relativity Theory; its exposition lies outside the scope of this review. However, QFT is the best framework for describing interactions among elementary particles; this research field being characterized by new symmetry properties and, sometimes, by asymmetry ones.

Elementary Particles and Fundamental Symmetries

As shown in the preceding sections, Relativity Theory introduces fundamental tools for apprehending new physical processes. These are essentially:

- Relativistic Conservation Laws, like the conservations of the total energy-momentum (including the mass) and of the total angular momentum (including the spin of the particles).

- Relativistic Covariance. It is a new principle which stipulates that the description of any quantum process must have the same analytical form, independently of the reference frame in which the process is observed; the best example being the computation of the transition probabilities between different quantum states.

- According to these new principles, E. Wigner wrote a paper devoted to the representations of the Lorentz Group [20] which constitutes a new breakthrough in Relativistic Quantum Physics (including QFT). He applied techniques of Group Theory to the Hilbert space describing a quantum system obeying Relativity principles; and he succeeded to derive two Relativistic Invariants (RI), independently of the internal dynamics of the physical system. The two RI are respectively the mass of the quantum system (already found since the introduction of the Minkowski 4- space) and its intrinsic angular momentum or its quantized spin.

An immediate consequence of these RI is related to the massless particles which move at the speed of light c like photons, gravitons and even neutrinos if their masses are neglected: the projection of the spin of a massless particle along its momentum or helicity, $\lambda = s \cdot p/|p|$, has only two values, $\lambda = \pm s$ for $s = 1/2, 1, 3/2, \cdots$. It could be seen that the massless quantum state with $\lambda = -s$ is the mirror image of the other state with $\lambda = +s$ and vice-versa. This intuitive symmetry will be tested experimentally and important features will be deduced.

- It is quite remarkable that Wigner's achievement completes the seminal work of Dirac: by means of the relativistic formulation of the quantum principles, the intrinsic angular momentum of a particle, or its spin, emerges directly. Spin is an essential ingredient for describing atoms, nuclei and particles; and it is worth noticing that, thanks to the existence of the spin, Pauli formulated his famous Exclusion Principle which explains the stability of atoms and molecules.

- Other important symmetries in Subatomic Physics are the Discrete

Symmetries whose experimental tests are still a challenge for Nuclear and Particle Physics. These are:

1) Charge Conjugation, C, which transforms any particle with charge q into its antiparticle with charge $^{-q}$ by keeping its mass and its spin unchanged (it is a consequence of the relavistic Dirac equation).

2) Parity, P, (or mirror symmetry) which transforms the vectors $r, p,$ and spin s into $-r, -p, s$.

3) Time Reversal, T, which transforms $r, p,$ and spin s into $r, -p, -s$, and exchanges both initial and final states.

It is usually argued that a quantum system is invariant by one of these three symmetries, $O_D = C, P,$ or T, if its hamiltonian H, which governs its evolution in time, keeps the same form after applying O_D on the physical system. Mathematically, it leads to $O_D H O_D^{-1} = H \Rightarrow$ commutator $[O_D, H] = 0$.

The two symmetries, P and T, are already used in NRQM, especially for establishing selection rules [21]; but their relativistic aspect becomes more crucial in Particle Physics which deals with energetic particles like electrons or neutrinos whose velocities are often very close to the speed of light c.

• In this context, sophisticated experiments have been performed in order to test the validity of these symmetries, as suggested by theorists T. D. Lee and C. N. Yang in 1956 in order to solve the "(θ, τ) puzzle" [22]. An historical experiment analyzing electron angular distributions from nuclear β decay [23], followed by another one studying weak decay of the muon, $\mu^- \rightarrow e^- \bar{\nu}_e \nu_\mu$ [24], proved in an undoubtful way that both Parity and Charge Conjugation are violated in weak interactions: whatever the initial decaying particle is, the neutrino is a left-handed particle, while the anti-neutrino is a right-handed one ([3]). But the combined operation, CP or PC, is still a good symmetry [25]. This hypothesis was true until the experimental discovery of Cronin et al. (1964) which shown the breakdown of CP symmetry in the neutral meson system, $\left(K^0 \bar{K}^0 \right)$, made out by a pair of particle-antiparticle evolving differently with time [26].

• After the failure of the Discrete Symmetries, a main question arose [27]:

Discrete symmetries being only approximate, is there any fundamental law which could relate a particle or a system of particles to their corresponding antiparticles?

The answer to this crucial question comes again from Relativistic QFT, thanks to the CPT theorem [28] which states the following: "Assuming Lorentz

Invariance of the interactions among particles, physical laws are invariant by the product of the three discrete symmetries C, P and T taken in any order: CPT, TPC, TCP, ...".

This theorem concerns any local interaction described in the Minkowski 4-space, like Strong, Electromagnetic and Weak interactions, and does not depend on the fact that any discrete symmetry is conserved or violated. It leads to important physical consequences which are still valid on the experimental side:

- The equality of mass of any particle (X) with its antiparticles $(\bar{X})$ [29].

- The total lifetimes of X and $\bar{X}$ are identical.

- The intrinsic magnetic moments of X and $\bar{X}$ are opposite in sign but they have the same absolute value.

- Inverse reaction with antiparticles:

$$a+b \rightarrow c+d \quad (1) \quad \text{and} \quad \bar{c}+\bar{d} \rightarrow \bar{a}+\bar{b} \quad (2)$$

Reaction (2) has the same probability of occurrence than reaction (1), even if symmetry CP or T is violated.

Quarks, Leptons and Gauge Fields

Gauge field theories and their applications to Particle Physics represent today the heart of Modern Physics and of all the activities turning around them, both theoretical and experimental. It requires a huge literature to describe them and, in this paragraph, emphasis will be put on the basic principles and the original ideas which guided their development.

- According to their physical properties (mass, charge, spin) and to their mutual interactions, particles are classified into specific families called multiplets. An evident example is the proton-neutron system which forms a doublet with regard to the strong interaction. Thus, it is said that the relativistic proton-neutron interaction is invariant by the symmetry group $SU(2)$.

After the discovery of many particle families, new quantum numbers arose like strangeness, charm, beauty..., and their classification become an urgent problem for Theoretical Physics. Despite their differences, particles present several common aspects and notably symmetry properties which lead to the use of Group Theory techniques.

Finally, two important families emerged: quarks and leptons, and their interactions are described by gauge fields.

Gauge Field

A gauge field is any field (classical or quantized) which remains invariant by a gauge transformation, like Classical Electrodynamics or QED. Let us consider the electromagnetic 4-potential $A = (A^\mu) = (V/c, A)$. If each component A^μ is transformed like: $A^\mu \rightarrow A'^\mu = A^\mu + \partial^\mu f$, f being any continuous function of (x), the electromagnetic tensor $F^{\mu\nu}$ (defined in sect.2) will remain unchanged.

This simple transformation (Abelian in the language of Group Theory) can be generalized to other fields, like nuclear field or quark field, by taking account of the "internal symmetry" which characterizes particles belonging to the same multiplet [30]. In this last case, the gauge transformation is called a non-Abelian one.

Quarks

Quarks [31] are the fundamental constituents of the Hadrons, heavy particles like Mesons and Baryons which are produced by strong interactions. When interacting among themselves, quarks exchange gluons (massless partcles like the photon) and this kind of interaction requires a new quantum number called color. The color is described by a relativistic gauge field theory which is similar to QED (describing electron and photon interactions). It is called Quantum Chromodynamics (QCD) [32].

All the quark properties are deduced from symmetry principles underlying the physical properties of the hadrons [33]. Their charges are fractions of the elementary electric charge e, and they can be classified into two groups, light quarks and heavy ones. They are respectively:

$$u(2/3), d(-1/3), s(-1/3), c(2/3), b(-1/3), t(2/3)$$

Number inside parenthesis indicates the charge of the quark in unit of e; the three first quarks, u, d and s, are the light ones while the three last ones are the heavy quarks.

Quarks have another physical property: they undergo electromagnetic and weak interactions like the hadrons in which they are confined. Thus, in the Electroweak Model of Glashow-Weinberg-Salam [34-36], quarks are classified into doublets, each doublet having an Up quark with charge $+2/3$ and a Down quark with charge $-1/3$.

Leptons

The word Lepton means initially a light particle, like the electron whose mass is approximately 1856 smaller than the proton mass. The neutrino, which is simultaneously produced with the electron in β decay, is a lepton too. Then, after the discovery of the muon, $\mu^{\pm}$, which has the same properties than the electron except its mass (the muon is 210 heavier than the electron), "lepton" usually indicates particles which are not sensitive to the strong interaction. Leptons are sensitive only to electromagnetic and weak interactions; and each lepton has its own quantum number called the "leptonic number".

At this level, particle physicists noticed a stringent similarity between Leptons and Quarks in the electroweak domain; it was the birth of the Standard Model which classifies all the fundamental particles into doublets according to their electroweak interactions [37].

The discovery of the third charged (and heavy!) lepton, the $\tau^{\pm}$, with its associated neutrino the v_τ [38,39], followed by those of the quarks b (beauty) and t (top), confirmed in a brilliant manner the exactness and the accuracy of the Standard Model (SM).

Needless to say that the SM is now a true theory. All its predictions have been verified experimentally: aside heavy quarks cited above, there were the discovery of the massive intermediate vector bosons, $W^{\pm}, Z^0$ (CERN 1983) and the one, more spectacular, of the Higgs boson (CERN, 2012-2013) whose field generates the mass for the fundamental constituents of matter.

CONCLUSIONS

Attempt has been made to give a large survey on the importance and intrication of Relativity Theory with Symmetry in the physical processes. From Classical to Modern Physics especially Particle Physics, numerous aspects of symmetry would not be understood even discovered without the principles of (Special) Relativity. Thanks to the wedding of Relativity and Quantum Mechanics and the inclusion of new symmetry principles, the resulting quantum fields get a predictive power like the Standard Model one. However, several fundamental problems remain unsolved:

Do magnetic monopoles exist in Nature? Their absence still indicates a dissymmetry between Electricity and Magnetism. Could modern Theoretical Physics realize a complete synthesis between General Relativity and Quantum Mechanics, as Dirac achieved eighty years ago? At a deep level of Quantum Mechanics, is there a real symmetry, or super symmetry, between Bosons and Fermions?

All these open problems require a large use of symmetry principles including sophisticated methods from Group Theory, which is nowadays visible with the impressive development of Mathematical Physics. On another side, experimental physics could answer and eventually solve all these passionate problems.

ACKNOWLEDGMENTS

The author would like to thank many colleagues and friends for stimulating discussions about symmetries in Physics. He is particularly very grateful to V. Morenas whose comments and remarks helped him in clarifying difficult problems.

REFERENCES

1. Curie, P. (1894) Symmetry in electric and magnetic phenomena (Sur les Symétries des phénomènes électriques et magnétiques). Journal de Physique, série III, t. III.

2. Weyl, H. (1964) Symmetry and mathematics (Symétries et Mathématiques Modernes). Flammarion.

3. Einstein, A. (1905) On the electrodynamics of moving bodies (English version). Annalen der Physik XVII.

4. Lorentz, A.H. (1904) Electromagnetic phenomena (English version). Proceedings of the Academy of Sciences of Amsterdam, 6.

5. Minkowski, H. (1908) Space and time. Address delivered at the 80th Assembly of German Natural Scientists and Physicians, Cologne.

6. Landau-Lifschitz (1994) The classical theory of fields. 4th Revised English Edition, Pergamon.

7. Einstein, A. (1905) Does the Inertia of a body depend upon its Energy content. Annalen der Physik XVII.

8. Einstein, A. (1922) The meaning of relativity. Lectures delivered at Princeton University, 5th Edition, Princeton University Press, Princeton.

9. Wigner, E. (1959) Group theory and its applications to the quantum mechanics of atomic spectra (English version). Academic Press Inc., New York.

10. Dirac, P.A.M. (1984) The principles of quantum mechanics. 4th Edition Revised.

11. Einstein, A. (1905) Point de vue heuristique concernant la production et la transformation de la Lumière (French version). Annalen der Physik, XVII, 132-148.http://dx.doi.org/10.1002/andp.19053220607

12. De Broglie, L. (1930) Introduction to wave mechanics (Introduction? L'étude de la M? canique Ondulatoire). Lectures given at the Henri

Poincare Institute, Masson Editions.

13. Messiah, A. (1995) Quantum Mechanics I (Mécanique Quantique. Tome I). Masson Editions.

14. Messiah, A. (1995) Quantum Mechanics II [Mécanique Quantique. Tome II). Masson Editions.

15. Dirac, P. (1928) Quantum Theory of the Electron I and II. Proceedings of Royal Society, London, A117, 610 and A118, 351. http://dx.doi.org/10.1098/rspa.1928.0056

16. Dirac, P. (1934) Théorie du Positron (published in French). Rapport du 7eme Conseil Solvay de Physique, Structure et Propriétés du Noyau Atomique, 20.

17. Anderson, C.D. (1933) The positive electron. Physical Review, 43, 491. http://dx.doi.org/10.1103/PhysRev.43.491

18. Cahn, R.N. and Goldhaber, G. (1989) The experimental foundations of particle physics. Cambridge University Press, Cambridge.

19. Schwinger, J. (1958) Selected papers on quantum electrodynamics. Dover Publications.

20. Wigner, E. (1939) On unitary representations of the inhomogenuous lorentz group. Annals of Mathematics, 40, 39. http://dx.doi.org/10.2307/1968551

21. Wigner, E. (1959) Group theory and its applications to the quantum mechanics of atomic spectra (English version). Academic Press Inc., New York. Chapters 18 and 26.

22. Lee, T.D. and Yang C.N. (1956) Question on parity conservation in weak interactions. Physical Review, 102, 290. http://dx.doi.org/10.1103/PhysRev.102.290

23. Wu, C.S. (1957) Experimental tests of parity conservation in beta decay. Physical Review, 105, 1413. http://dx.doi.org/10.1103/PhysRev.105.1413

24. Garwin, R.L., Lederman, L.M. and Weinrich, M. (1957) Observation of the failure of conservation of parity and charge conjugation in meson decays. Physical Review, 105, 1415. http://dx.doi.org/10.1103/PhysRev.105.1415

25. Landau, L.D. (1957) Conservation laws in weak interactions. Journal of Experimental and Theoretical Physics, 32, 405-406.

26. Christenson, J.H. and Cronin, J.W. (1964) Evidence for the 2π decay of the meson. Physical Review Letters, 13, 138. http://dx.doi.org/10.1103/PhysRevLett.13.138

27. Wigner, E. (1964) Symmetry and conservation laws. Physics Today, 17,

34-40.http://dx.doi.org/10.1063/1.3051467

28. Several authors have postulated the CPT theorem by similar approaches. The pioneering one was the paper of G. Lüders (1957) Proof of the TCP theorem. Annals of Physics, 2, 1-15.

29. Lee, T.D., Oehme, R. and Yang, C.N. (1957) Remarks on possible noninvariance under time reversal and charge conjugation. Physical Review, 106, 340.http://dx.doi.org/10.1103/PhysRev.106.340

30. Yang, C.N. and Mills, R.L. (1954) Conservation of isotopic spin and isotopic gauge invariance. Physical Review, 96, 191. http://dx.doi.org/10.1103/PhysRev.96.191

31. Gellmann, M. (2013) The quark model. International Journal of Modern Physics A, 28, 1330016.

32. Greenberg, O.W. (1964) Spin and unitary-spin independence in a paraquark model of baryons and mesons. Physical Review Letters, 13, 598.

33. Lichtenberg, D.B. (1970) Unitary symmetry and elementary particles. Academic Press, New-York.

34. Weinberg S. (1980) Conceptual foundations of the unified theory of weak and electromagnetic interactions. Reviews of Modern Physics, 52, 515.

35. Abdus, S. (1980) Gauge unification of fundamental forces. Reviews of Modern Physics, 52, 525.

36. Glashow, S.L. (1980) Towards a unified theory: Threads in tapestry. Reviews of Modern Physics, 52, 539.

37. For a clear and concise introduction to the Standard Model, one can consult the excellent book of C. Quigg (1983) Gauge Theories of the Strong, Weak and Electromagnetic Interactions. Frontiers in Physics.

38. Perl, M.L. (1975) Evidence for anomalous lepton production in e^+e^- annihilation. Physical Review Letters, 35, 1489.

39. Brandelik, R. (1978) Measurement of Tau decay modes and a precise determination of the mass. Physical Letter B, 73, 109.

NOTES

*The author dedicates this paper to the memory of his mother, Elise Ajaltouni-Haddad (1922-2012), for her love and her affection.

$1^{(A^\mu)}$ is the four-vector potential given by $(V/c, A)$ and ∂_μ is the partial derivation according to the component x^μ.

2^{ν} is the velocity of the massive particle which is identified with the group velocity of the wave, while V is the phase velocity which could be greater than c, h is the Planck constant and $\gamma = 1/\sqrt{1-v^2/c^2}$.

[3]Recalling that helicity is given by $\lambda = \dfrac{s \cdot p}{p} = \pm 1/2$, only neutrinos with $\lambda = -1/2$ called left neutrino ν_L exists, while only anti-neutrinos with $\lambda = +1/2$ called right anti-neutrino $\bar{\nu}_R$ exists in Nature. The transformation allowing the transition from ν_L to $\bar{\nu}_R$ is the product of C and P, or CP Symmetr

Chapter 5

THE ORIGIN OF PARTICLES

Edwin Zong

Medicine Department Oasis Medical Group, Bakersfield, USA

ABSTRACT

The emulsion exists ubiquitously. The mass loves to be segregated based on its inherited similarity. Since everything is made from atoms/nucleus/electrons, it is not surprised to know that those little creatures share same flavor with their bigger brothers e.g. oil and water. In a lab, the nuclear emulsions have been well studied and used to investigate fast charged particles like nucleons or mesons [1 [2. Similarly, the author further reasons that the phenomenon of emulsions may very well exist for sub-nucleon particle, more specifically, photons! The photons distinguish themselves with various wave length and frequencies. The photons of similar wave length/frequency maybe segregated in the emulsion soup of electron and/or dissident photons with different wave length/frequencies! The big bang Nucleosynthesis is a well-established fact in our physics [3. The author has subsequently linked gamma ray photons to our big bang [4. Based on the conservation law of mass, our galaxies must come from particles generated from a big bang, rather than from nothing. More specifically, our nucleon and sub-nuclear particles are all from photons, because the photons are the only option for linking our mass to a big bang of high energy particles/λ ray photons! The dark matter is a congregation of photons forming immediately after a big bang! The visible galaxies appear later. It is the emulsion powers that drive photons to be segregated base on their similar wavelength/frequencies. The bundled photons are building blocks for forming protons, neutrons, black matter and so on. Before photons are coupled /bundled, they manifest no gravity effects; therefore, the emulsion force is the main force behind segregation. Once photons are coupled, they may manifest gravity effects [4 and gain ability to attract electrons which further stabilize the structure of bundled photons or neutrons [5! What confuses people is our perception of photon – how storming particles with speed of light is able to "slow down" and become part of emulsion soup. Because the mass or quarks sit there orbiting rather than storm with speed of light. The missing link is now founded; it is "photon bonding/coupling"!–a recent remarkable discovery called Rydberg blockade by coaxing photons into bonding together to form "molecules" – a state of matter by a group led by Harvard Professor of Physics Mikhail Lukin and MIT Professor of Physics Vladan Vuletic [6. The primary objective of this study is to uncover the origin of all particles and map the mass evolution in the discipline of basic physics.

INTRODUCTION

The linkage of a big bang to high energy particle or λ ray photons is crucial for fundamental particle study since our tools are often limited when we try to dissectour tiny little bit creatures named particles! However, all roads lead to Rome. The correct understanding of our universe origin will help us understand our particle world! If our big bang is truly a "high energy particle" blast, based on conservation law of mass, we can be 100% confident in pining down photons are the most basic building block for all particles and so on! In other words, all particles indeed have same single origin!

The particle world holds the answer for everything. Unfortunately, limited by our digging tools, everyone tumbles at quark level. Without forming any new theorybeyond basic physics! The author can assure everyone that quark is originated from "settled" photons! This conclusion is supported by big bang theory, because all masses are the products from a big bang which is a high energy particle blast.

Unfortunately, no matter how weak a laser/light source we use, we are far from isolating a single photon or a single proton in our lab which demands absolute temperature "near" zero! It is that reason we cannot pin down anything beyond quark in experiment. However, the tremendous "slowing down" of flying photons (uncoupled photons) in extreme cold medium supports the hypothesis that photons can be emulsified! Other than emulsion power, there is no other candidate force that can be "remotely possible" used to explain the phenomenon of photon bonding [6! The emulsion appears to exist ubiquitously; the only exception is void space. Nothing can be emulsified against void. Fortunately, our universe is never a void place [4! In 1964, cosmic microwave background (CMB) was discovered by Arno Penzias and Robert Wilson [7! The CMB provides a perfect universal emulsifying background agent in space!

Biological photosynthesis is another common natural phenomenon where photons are captured and incorporated in mass. The maximum amount of photosynthesis per incident unit of energy is at a wavelength, around 650 nm (deep red). If we limit the supply of quantity of such red solar lights on chlorophyll (other words, not to overwhelm it), the lights of study will be disappeared entirely right in front of our eyes! Unfortunately, the widely used term "light absorption" is a very ambiguous word that covers the most remarkable process in particle physics! The photosynthesis of apple tree leaves contributes to apple tree weight/gravity [4! Apparently, the solar photons of wavelength 650 nm cannot be bounded at nuclear level or proton/neutron level of apple tree, otherwise, the different atoms will be created if proton/neutron mass number is changed by photosynthesis! The equation CO_2 + H_2O+Solar photons→$C_6H_{12}O_6$ (glucose)+O_2 tells us there is no changing of atoms during photosynthesis. The author, therefore, can safely deduce that solar lights are bonded at sub-particle level! No matter how fancy the particles act, they are all made from same origin foundation particle- the photons! There is a "zoo" of particles, which may simply correspond to light with "zoo" of different wavelength/frequencies before such photons are captured. The different wavelength/frequencies a photon once owned, cannot be disappear based on the conservation law of mass and energy, therefore, it affords "different character" possible for particles!

In any soup of distinguished different type of mediums, the photons of similar wave length/frequencies will be congregated together through a well-known process called emulsion! Because other than emulsion power, there is no other candidate force that can be "remotely possible"used to explain the phenomenon of photon bonding [6!

To stable the "bonded" photon structure, also known as Neutrinos, we need electrons! On the other hand, electrons are stabled by "bonded" photon structure as well. The stable effects between them are mutual! The author's view is supported by nuclear reaction, in which, if photon strike/disturb the neutrinos, the electrons will be excited. It is also evidenced in "Bond Softening" experiment [8!

SOLAR PHOTONS, PHOTON-BONDING AND PHOTON EMULSION

We all love our sun, because it tans us, but little we know, the sun also serves a greatest particle lab for us free of charge! There are tremendous amount of photons coming out of sun every second, but the sun is not made of photons! The sun is mainly made of hydrogen along with other elements but photons [9! So, where are solar photons from? The hydrogen! According to conservation of mass, no matter can be created or destroyed, solar photons cannot come from nothing, the photons are from hydrogen, more specifically, their nucleons! This is a destructive way to prove photons are the origin for all particles! Of course, not all solar hydrogen will puff up becoming solar photons due to its enormous size with various physical conditions in the sun. nonetheless, this sun's business is sufficient to tell us that photons can be generated from its nucleons! The author can also present you a constructive way to prove photons are the origin for all particles!

Photon Challenge

All stimulus need to be translated into electric current for human brain comprehension. The vision is no exception. Here is an example for our seeing a yellow apple. For a ripened yellow apple, wavelengths of about 570 to 580 nanometers bounce back. These are the wavelengths of yellow light. When you look at a yellow apple, the wavelengths of reflected light determine what color you may see. The light waves reflect off from the yellow apple and hit the light-sensitive retina at the back of your eye.

There is an inherited barrier for us to manipulate photons down to a single level. If that single photon is not reflected away from experiment mediums, and hit in human eye, no one would see it. It is also highly unlikely for human eye detect one single photon because one single photon unlikely provokes light threshold and sparks electric current in human eye cells. For us to spot a light, not only we need our lens to increase concentration of photons, but also our eye has animpressive high light threshold, considering 6 to 7 million cones concentrated on a 0.3 millimeter spot on the retina. It requires certain quantity of photons to incite a reaction.

In addition, in a lab chamber with one photon or single proton, it demands absolute temperature of "near" zero! Otherwise, it is a contaminated lab chamber [10. The author is afraid to say that such demanding is beyond human technology that will ever reach.

It is a formidable barrier for us to manipulate a single photon or single proton in an "absolute" void chamber, but it is not that hard for us to look into space and deduce that our particles are all from one single origin, that is photon!

Photon-bonding or Photosynthesis in Space

The object of biochemistry is part of universe which is subject to the same laws of physics. The author would be surprised if the objects of chemistry follow different laws of physics. Photosynthesis is a well know biological process where "uncoupled" photons are captured by plants. The word of "capture" is a very vague term that conceals a most remarkable "particle" process in physics. Apparently, the "uncoupled" photons must be bonded for new molecule "glucose" forming! Otherwise, the storming "un-coupled or un-bonded" solar photon will go thru or reflex from leaf without "captured".

The photosynthesis apparently is not limited to solar photons! There are some types of fungi, called radiotrophic fungi, are able to use melanin as a photosynthetic pigment that enables them to capture gamma rays [13 and harness the energy for growth [14. As author always believes, the nature is never penny pinching in providing us facts that help us to discover the laws of nature. The λ ray photosynthesis is extremely remarkable discovery for two reasons! 1. At the early stage of earth, λ ray must be abundantly available; the organisms apparently are very adaptive that can utilize any types of photons. 2. λ ray photosynthesis is real.

The universe space is never a void place [4. Here comes the cosmic microwave background (CMB) which is the thermal radiation left over from the "Big Bang". It is the oldest light in the current observed universe. Moreover, the microwave photons are un-coupled and exist as free flying ray! It serves one of forever emulsion agent in space! In a sense, any mass/matter possesses different character of physics will subject to segregation emulsion in a "CMB" soup!

Our current cosmos Gamma ray (not all of them) is from a big bang! It is partially supported by fact that the energies of gamma rays from astronomical sources range to over 10 TeV. That energy is far too large to result from radioactive decay [11. some cosmos gamma ray are from pulsar and black hole though. However, gamma-ray pulsars and rare occurrence of gamma ray burst from black hole are very rare events in the universe. It is evidenced by recent observation: there have been only about one hundred gamma-ray pulsars identified out of about 1800 known pulsars [12. Base on observation, the sources of most GRBs are billions of light years away from Earth, implying that the explosions are both extremely energetic (a typical burst releases as much energy in a few seconds as the Sun will in its entire 10-billion-year lifetime) and extremely rare (a few per galaxy per million years). Given the facts that the universe space is infinite big and it is full of diffuse gamma radiation, it can be reasoned that the pulsars and black holes are not sufficient source for such extensive universe background gamma radiation. Secondly, the radioactive decay is not a right candidate power house for cosmos γ-ray background either. The only option left here is a big bang. The big bang provides most of those cosmos γ-ray today. The remaining cosmos γ-ray is provided by big bang indirectly.

Apparently, the initial blast of gamma ray photons from a big bang will penetrate any mass that exist during a big bang explosion. For any gamma ray photons to be trapped or coupled at any sub-nuclear level, those storming photons need to be "slow down" first! Otherwise, the emulsion will not work with storming rays at speed of light! The missing link is how lights can be slow down. Now the puzzle is finally solved! At couple degrees above absolute temperature zero, the lights appear "slow down" significantly where ubiquitous emulsion stands for a chance to work! Our universe space background temperature is 2.7 Kelvin. It might go further down when our region of universe is ripe for next big bang, however our background temperature will never reach absolute temperature zero [4.

The universe is a complete and enclosed system, because there is nothing existing outside universe which may interferes universe matters/energy. The conservation of mass and energy perfectly apply universe as a unit. We can safely say that universe energy and mass are constant. $E_{bigbang} = E_k + E_p$

The E_k is the sum of kinetic energy presents in the current universe. E_p is the sum of positional energy presents in the current universe. The big bang is a greatest event happened in the universe; it is an almost pure kinetic energy show. Soon after a big blast, some of its kinetic energy starts to transfer itself to the positional energy (γ ray nucleosynthesis/ photosynthesis). Some γ ray photosynthesis trapped at subatomic stage and evolves into dark matter (significant positional energy, little kinetic energy). Some γ ray nucleosynthesis goes on to form galaxies and us. Some γ ray remains free standing status and becomes cosmos background ray [4. The vast majority of λ ray fades into cosmic microwave background [16.

The light photons in a big bang blast provide building block for everything that possess positional energy include you and me. Apparently, light photons are not equal to proton or neutrons. But those particles are all originated from photons based on deductions of reasoning: 1.The big bang is the blast of high energy particles/λ ray photonsbased on observation [4 2. All Mass/Matters/neutrons are from high energy photons/big bang based on conservation law of mass.

Electrons and Friction

Electron apparently helps stabilize particle bonding and/or photon bonding.

As we all know, the emulsion requires initial energy input which is the big bang. As time passes, the big bang kinetic energy declines, the emulsion will revert to the stable state of the phases comprising the emulsions (e.g. the congregation of photons, the separation of neutrons and electrons in space). In dark matter, the initial segregation of electrons and neutrons actually destabilizes "newly formed" dark matter physics. Fortunately, such destabilization is counter-balanced by gravitation among emulsion agents when they grow into a substantial size! In a sense, all dark matter will have rich electrons around (gravity effects) which in turn help stabilize their infrastructure of neutrons [15 or bundled λ ray photons!

The halo of rich electric activities (e.g. x-ray), therefore, will be expected from any dark matter of substantial size! The destabilization occasionally affects neutrons which may unleash its bonding λ photons - excess λ photon activities!

The halo of dark matter (electron clouds) will be expected in space and experiment as well! Such halo is a norm for any black hole in space since black hole is made from dark matters! The halo and dark matter are kept together because of gravity effects between them! In addition, the halo (electron clouds) may also serve "friction factor" when multiple black hole merge. It is evidenced by recent discovery of a "lagging effect!" black hole in the merger of multiple black holes due to said frictions, where merging of first two black holes likely makes "halo" or electron clouds much thicker-creating greater friction for the thirdblack hole to join! Those frictions eventually will be overcome by tremendous gravity among black hole!

In organic world, the photosynthesis is very common phenomenon, the uncoupled/unbounded photon hit chlorophyll and accessory pigments such as carotenoids and phycobilins, chemical bonding "softening" happens, electron subsequently released, meanwhile, plant molecule potential energy curves become distorted. The photons apparently join the particles of receiving plant molecule. It is not only supported by energy curve distortion observed from experiment, but also evidenced by weight gaining/gravity gaining from growing plants which capture large quantity of photons through their growth. Other than certain wavelengths of solar photons are preferred by most plants, we also know radiotrophic fungi, which are able to use melanin as a photosynthetic pigment that enables them to capture gamma rays [13 and harness the energy for growth [14. γ-ray is probably abundance in the early stage of earth while atmosphere has not developed. The organic lives must try every possible way to capture light photons to power their body!

In the space, the dark matter is stable mass rich in "coupled" or "settled" high energy particles/λ photons, its stabilization factors/electrons must be rich as well.

The author reasonably expect more activities happen among electrons within or surround dark matters, which may strike and trigger X ray. The author's view is supported by X-ray emission observed from center of galaxies [18. On the other hand, fewer events gain enough power that "directly" disturb coupled λ photons, which may be unleashed and manifest as λ ray activities in dark matters. The findings of λ ray burst or excessive λ ray photons in the black hole may indicate the transparency character of dark matter for λ ray or non-absorption of λ wavelength "uncoupled" photons! Which also indicates that dark matters "coupled" λ photons may reach its saturated level, leaving fewer rooms for "extra" λ photons bonding.

Dark Matter Experiment

In the space, the continuing of merging galaxies leave bigger and bigger space blobs with extremely low temperature which is a cycling process of preparing next big bang. Once a big bang occurs, the high energy particles of big bang most likely encounter many regions of extremely low temperature in space where dark matter start to form. The Lukin and Vuletic's experiment solved a critical part of puzzle how "light speed" photons "slow down" affording emulsion process to do its job totransfer their kinetic energy to positional energy! The author has deduced dark matter is made from γ-ray photon-bonding or photosynthesis. Since γ-ray photosynthesis is not total alien to earthlings, we do have difficult time to understand how γ-ray photosynthesis occurs in the space. The author is very pleased to find out recent remarkable discovery called Rydberg blockade by coaxing photons into bonding together to form "molecules" – a state of matter by a group led by Harvard Professor of Physics Mikhail Lukin and MIT Professor of Physics Vladan Vuletic.

The big bang blast is most likely high energy particle e.g. λ ray photons, it is not electrons. In the immediate post bang "soup", the "space emulsion" makes λ photons segregation possible in the soup of microwave inferred photons and electron clouds. The initial space emulsion agents may not be structured as stable as the ones in later stage of universe evolution. The more stable structure develops when multiple nuclear decays happen in a little later stage of mass evolution!

The emulsion force that bundles "frozen" λ photons against microwave inferred photons and electrons is weak, therefore, nickname those little creatures as weakly interacting massive particles is deemed appropriate!

In the lab, Nuclear emulsion was first discovered by Marietta Blau and Hertha Wambacher In 1937. The author reasonably believes the dark matter can be formed in a lab setting where appropriate medium or photographic plate available as background/emulsion soup. The targeting photons must be λ ray photons! The temperature must mimic deep space temperature (2.7 kelvin or lower). The author also reasonably believes the electrons must exist in the medium for stabilizing newly formed dark matter in lab!

METHUSELAH STAR HD 140283

The author applaud international collaboration for the Dark Energy Survey, if basic physics works for deep space, during big bang preparing stage, the map should demonstrate the "greater and greater" concentration of "matters or dark matters" as mass evolves or timing by humans, a phenomenon is similar to "oil water emulsion". Meantime, we will see larger and larger space blobs forming with little mass and extremely low temperature inside. It is evidenced by newly discovered primordial blob which formed in prior circle of big bang preparing stage. Since the space is never void during or before any big bang, it will not be surprise to see some "older" celestial mass which is older than a current observed universe! It is supported by discovery of Methuselah star, HD 140283 which is "older" than our current universe! It is not paradox, it is another perfect evidence supports author's universe model- one universe, endless cycles in a realm of gravity and emulsion [10!

While universe expansion is accelerated, the frequency of merging galaxies/black hole may drop. The larger space blobs means greater distance to overcome for galaxies/black hole to meet. However, the size of galaxies/black hole will be significant larger than the ones at early stage of universe development! The larger size of galaxies/black hole make universe expanding acceleration possible due to increasing gravity power! The mature stage of galaxies may be significantly dimmer after large amounts of big bang photons ("visible" uncoupled ray) are bundled into positional energy to form galaxies mass ("coupled" photons), though.

Emulsion and Gravitation Revision

The emulsion plays a key role for universe galaxies formation and nucleon formation. The gravity power is much more significant comparing to emulsion power, especially in late stage of universe development, when larger and larger galaxies/black holes formed! The emulsion force sometimes works along with gravitation power, sometimes it will work against gravitation power. The emulsion power maybe Negligible when Newton calculated the apple velocity with gravity in his garden. Apparently, the apple is subject to emulsion force as well; because the physics character of apple is differ from air. However, it is negligible because the emulsion force is trivial!

The story might not be true, when we look at a star/planet with trillion and trillion size/mass of Newton's apple, the emulsion force between a gigantic star against cosmic background photons may not be trivial! The modified gravitation may become necessary. Same rule may also apply for calculating orbing status of a gigantic celestial mass!

The universe is never a place of completely "void". It is filled up with matters of two kinds – "uncoupled" and "coupled" mass at different concentration in space. The "uncoupled" mass e.g. cosmic ray corresponding to mostly "kinetic energy" and electrons; the "coupled" mass corresponding to mostly "positional energy" , those "coupled" and "uncoupled" mass serve distinguish different nature of mediums where "emulsion" comes to play. Therefore, the gathering of bigger and bigger mass is driven by two forces! – the gravity and emulsion. The author believe emulsion is most likely playing a bigger role for dark matter forming! rather than a result of simple collision of photons!

Matter, Anti-Matter, Space-time, Gravitational Wave

The author has always felt that basic physics has provided us remarkable understanding of mechanics behind everything on earth. To abandon them, the author thinks that we need a good reason. We all loveour academic freedom, but we are dedicated to seek truth as scientists and open for scientific scrutiny or debate!

It was 1756, Mikhail Lomonosov discovered the law of mass conservation by experiments, and came to the conclusion that phlogiston theory is incorrect. The concept of mass conservation is widely used in many fields (limited in scientific aspect of human life, of course) such as chemistry, mechanics, and fluid dynamics. However, the confusion of mass law persists deep into modern time! As author described earlier, when uncoupled/visible photons become bonded/coupled, "visible" photons will disappear in front of our eyes, but it doesn't mean they are vanished! The perception of "canceling" effects among matter and anti-matter implies "vanish". Such perception may validate a magician's dream work or trigger a mass worship, but it serves an intellectual black hole in physics.

Any particles that are coupled /congregated will manifest gravitational effects [4. The simple example is an apple, thru photosynthesis; the big apple that contains large quantities of coupled photons along with new electrons may manifest gravitation effects! The solar light can be, therefore, considered as gravitational wave. However, the gravitational wave is misleading concept because it implies that there are non-gravitational wave/particles exist. As I described earlier, all particles include electrons and photons can manifest gravitation of significance! if they are coupled/congregated! There are no such things eligible for non-gravitational wave!

Space is just void where matter floats, and time is a history record of progress of matter evolution. The Space-time is not matter; it can never be "rubbed" or "rippled"!

Emulsion is a well-studied natural phenomenon in basic physics. The common sense of earthlings, however, tells the author that the space emulsion force may modify gravity force in space or anywhere, where stars, planets, black holes and /or particles are exposed to "non-similar" matter in a non-void space -emulsion soup!

The earth is part of universe! The law of physics works for earth; it will work for the rest of universe! And vice versa! The author is afraid to see that great thoughts of "icon of genius" might munch away many generations of bright mind (limited in science, of course!).

Nucleosynthesis, WIMPs, Radiation

The big bang Nucleosynthesis starts with merging /coupling of big blast high energy particles/photons in the soup of electrons and Cosmic Microwave background(initial big blast kinetic energy along with emulsion force)→protons and neutrons/primordial nucleons formation (some big blast kinetic energy transfer to positional or orbiting energy also known as gravity) → the "initial" or unstable nucleus decay/radioactive decay → dark matter formation/ stabilization + λ ray bursts → mass gathering due to dark mater gravity → visible galaxy formation.

The dark matter may go thru "initial formation" then "stabilization" with radioactive decay process:

$$\overline{V}e + p+ \rightarrow n0 + e+$$ (Cownan-Reines neutrino experiment) dictates beta negative decay reacted with protons to produce neutrons! and positrons which quickly combine with electrons, releasing gamma rays [17!

The author agrees with the increasing consensus among scientific community: Dark Matter is mass of neutrons with electron cloud that stabilize neutrons. The stable dark matter/black hole has signature "x" ray or "λ" photon excess. The initial "unstable" dark matter/black hole may have phenomenal λ ray burst! The λ ray burst, however, has dramatically reduced when universe progress to a more stable stage! So is the rate of galaxies/dark matter merge (decreasing), while size of galaxies/dark matters, however, are increasingly becoming gigantic when space emulsion force and gravity power continue to drive our current galaxies' merge.

The emulsion force that bundles λ photons against microwave inferred photons and electrons is weak, therefore, nickname those little creatures as weakly interacting massive particles is deemed appropriate.

SPACE LABORATORY EXPLORATION AND COMPUTER SIMULATION

There are two ways to learn flying. The airplane flight simulator or a real plane will all serve its purpose. Once we understand the makeup of all mass and the circle of universe, we will be able to quantify the possibilities by using computer to simulate the foreign stars/planets/galaxies/big bangs and so on. We can use computer to simulate all possibilities of organic matter evolution in different environment/planets. Instead of reaching out to visit aliens, we can create aliens in our big screen and watch them from our living room, a favorite way for travelling to many "home-bounded" tourists in today's world!

Our lab, however, should be able to create miniature version of pre-big bang dark matter mass, big bang, dark matter formation, galaxy formation etc.

RESULTS ANALYSIS

The emulsion is ubiquitously phenomenon, the mass love to segregate based on its inherited similarity. Since everything is made from atoms/nucleus/electrons, it is not surprised to know that those little creatures share same flavor as their bigger brothers e.g. oil and water. In a lab, Nuclear emulsions have been well studied and used to investigate fast charged particles like nucleons or mesons. Follow the same foot step, the author further reason that the emulsions also exist for sub-nucleon particle, more specifically, photons! The photons distinguish themselves with various wave length and frequency. The similar wave length/frequency photons may very well segregate in the emulsion broth of electron or photons with different wave length/frequencies!

The particle emulsions are only inches away from nuclear emulsions! Once we master particle emulsions, we will be able to produce dark matter in our lab! The practical application of making dark matter is enormous! Not only it will help us understand the origin of all particles, the origin of universe, but also it will re-shape our medicine, energy industry and military operation!

The universe is a complete and enclosed system, because there is nothing existing outside universe which may interferes universe matters/energy. The conservation of mass and energy perfectly apply un/iverse as a unit. We can safely say that universe energy and mass are constant. $\mathbf{E}_{bigbang} = \mathbf{E}_k + \mathbf{E}_p$

The $\mathbf{E}_k$ is the sum of kinetic energy presents in the current universe. $\mathbf{E}_p$ is the sum of positional energy presents in the current universe. The big bang is a greatest event happened in the universe; it is an almost pure kinetic energy show. Soon after a big blast, some of its kinetic energy starts to transfer itself to the positional energy (γ ray nucleosynthesis/ photosynthesis). It serves a practical basis for our reasoning that all particles are from same origin- photons! It is this reason, therefore, the author recommend use photons as emulsion agent in particle emulsion/dark matter experiment!

DISCUSSION

The mechanic of manufacturing all neutrons from photons is "emulsion"! The un-coupled photons/ray do not have gravitation effect, therefore, gravitation does not play a role in initial photon bonding! Once photon bonding occurs thru emulsion segregation, the gravity effects appear, bonding/ coupled photons, therefore, attract electrons around which will further stabilize photon bonding! The striking of a new photon on binding photons will destabilize the harmony between stabilized coupled photons and their mates-electrons. The electrons will be excited and jump, which further destabilize "coupled photon" until one photon will be unleashed from its bonding and fly out as a single ray/photon [17!

The traditional way of dissect particles may encounter forbidding technical difficulty by isolating single photon or proton for experiment. The recent discovery of photon bonding phenomenon in inorganic lab afford a possibility for making "photon emulsion" in human lab, furthermore, "producing our own dark matter" will be within our reach with this approach! We may see a "zoo" of particles, but they are most likely originated from photons based on universe observation in author's universe model.

Once we understand the fundamental makeup of all mass and the circle of universe; we can be"100%" confident to pin down all particles tosingle source-photons! We shall be able to quantify the possibilities by using computer to simulate the foreign stars/planets/galaxies/big bangs and so on. We can use computer to simulate all possibilities of organic matter evolution in different/exotic environment/planets.

The author promotes computer space exploration along with physical exploration in deep space.

REFERENCES

1.   G. Occhialini, C. F. Powell, Nuclear Disintegrations Produced by Slow Charged Particles of Small Mass, Nature 159, 186–190 & 160, 453–456, 1947.

2.   Nuclear emulsions by Ilford Harman Techonology Limited. December 2005.

3.   Doglov, A. D. "Big Bang Nucleosynthesis." Nucl. Phys. Proc. Suppl. (2002): 137-43. ArXiv. 17 Jan. 2002.

4. Edwin Zong, The Real Universe, Journal of Nuclear and Particle Physics, Vol. 5 No. 2, 2015, pp. 21-29. doi: 10.5923/j.jnpp.20150502.01.

5. Luis W. Alvarez, W. Peter Trower (1987). "Chapter 3: K-Electron Capture by Nuclei (with the commentary of Emilio Segré).

6. M. Lukin Harvard and V. Vuletic MIT Nature Sep 25, 2013.

7. Penzias, A. A.; Wilson, R. W. (1965). "A Measurement of Excess Antenna Temperature at 4080 Mc/s". The Astrophysical Journal 142 (1): 419–421.

8. P.H. Bucksbaum, A. Zavriyev, H.G. Muller and D.W. Schumacher "Softening of the H2+ molecular bond in intense laser fields" Phys. Rev. Lett. 64 1883 (1990).

9. Haubold, H. J.; Mathai, A. M. (1994). "Solar Nuclear Energy Generation & The Chlorine Solar Neutrino Experiment". AIP Conference Proceedings 320: 102. arXiv:astro-ph/9405040. Bibcode:1995AIPC..320..102H. doi:10.1063/1.47009.

10. Edwin Zong, One Universe, Endless Cycles, Science Research. Vol. 2, No. 5, 2014, pp. 105-110. doi: 10.11648/j.sr.20140205.15.

11. Aharonian, F.; Akhperjanian, A.; Barrio, J.; Bernlohr, K.; Borst, H.; Bojahr, H.; Bolz, O.; Contreras, J.; Cortina, J.; Denninghoff, S.; Fonseca, V.; Gonzalez, J.; Gotting, N.; Heinzelmann, G.; Hermann, G.; Heusler, A.; Hofmann, W.; Horns, D.; Iserlohe, C.; Ibarra, A.; Jung, I.; Kankanyan, R.; Kestel, M.; Kettler, J.; Kohnle, A.; Konopelko, A.; Kornmeyer, H.; Kranich, D.; Krawczynski, H.; Lampeitl, H. (2001). "The TeV Energy Spectrum of Markarian 501 Measured with the Stereoscopic Telescope System of HEGRA during 1998 and 1999". The Astrophysical Journal 546 (2): 898. Bibcode: 2001ApJ...546..898A. doi:10.1086/ 318321.

12. NASA'S Fermi Telescope Unveils a Dozen New Pulsars http://www.nasa.gov/mission_pages/GLAST/news/dozen_pulsars.html.   Cosmos Online – New Kind of pulsar discovered (http://www.cosmosmagazine.com/news/2260/new-kind-pulsar-discovered).

13. Castelvecchi, Davide (May 26, 2007). "Dark Power: Pigment seems to put radiation to good use". Science News 171 (21): 325. doi:10.1002/ scin.2007.5591712106.

14. Dadachova E, Bryan RA, Huang X et al. (2007). "Ionizing radiation changes the electronic properties of melanin and enhances the growth of melanized fungi". Plos One 2 (5): e457. doi:10.1371/journal.pone.0000457. PMC 1866175. PMID 17520016.

15. LARS BILDSTEN AND ANDREW CUMMINGHydrogen Electron Capture in Accreting Neutron Stars and the Resulting g-Mode Oscillation

Spectrum THE ASTROPHYSICAL JOURNAL, 506:842–862, 1998 October 20.

16. Durham, Frank; Purrington, Robert D. (1983). Frame of the universe: a history of physical cosmology. Columbia University Press. pp. 193–209. ISBN 0-231-05393-2.

17. C. L Cowan Jr., F. Reines, F. B. Harrison, H. W. Kruse, A. D McGuire (July 20, 1956). "Detection of the Free Neutrino: a Confirmation". Science 124 (3212): 103–4. Bibcode: 1956Sci...124..103C. doi:10.1126/science.124.3212.103. PMID 17796274.

18. Tauris & van den Heuvel (2006), "Formation and evolution of compact stellar X-ray sources", In: Compact stellar X-ray sources. Edited by Walter Lewin & Michiel van der Klis. Cambridge Astrophysics Series, p.623-665, DOI: 10.2277/0521826594.

Chapter 6

THE UNIFICATION OF PARTICLE AND ENERGY

Edwin Zong

Oasis Medical Group Inc., Bakersfield Ca USA

ABSTRACT

The electron (e^-) and a positron (e^+) are the decay products from neutron thru double beta decay. Cooking these two electrons of opposite charge/gender will break down to the gamma rays photons! The photon is generally accepted as electric charge neutral particle, so does the neutrino which happens to be the product from that neutron beta decay. The reverse of beta decay supports author's postulation which gamma rays are the nature of our big bang! Few will argue photons are electromagnetic/radiation power. However, many deny photon's mass status. In this paper, the author will assure you that photon is mass, which also happens to serve the foundation for all gravitational power/ positional energy that exists today after a big bang. Furthermore, the author will also assure you that neutrino is the dark matter that floats in the space chilling with electrons ($^-$ and $^+$). Again, the author assures you that the famous gamma ray burst is just cosmic-sized of our well-studied experiment in our lab: $e^- + e^+ \rightarrow \gamma + \gamma$! In the deep space, the reverse of beta decay happens immediately after a big bang! The γ ray photon forms electron and neutrino first. The neutrino is our evasive dark matter! The reverse of beta decay also supports the author's postulation that photons are mass as well! Our current universe's particles and forces (detectable or undetectable) are all derived from a big bang's photons. The primary objective of this study is to uncover the common origin for both mass and force in the universe along with their evolutionary path in the discipline of Newtonian physics.

INTRODUCTION

In medicine, we often choose to use Positron emission tomography (PET) scan to detect "highly active" cancer cells. The highly active molecule chosen for PET tracer is fluoro-deoxy-glucose (FDG), an analogue of glucose. The high concentration of tracer/fluoro-deoxy-glucose indicates highly metabolic activity which corresponds to the highly regional glucose uptake-cancer cells. It is the standard of medical care for using this tracer to map the cancer metastasis. The PET scan detects pairs of gamma ray photons emitted indirectly by a positron-emitting radionuclide (tracer), which is introduced into the body such as FDG. [1] The current physics equation for the process is as follows: e^- + $e^+ \rightarrow \gamma + \gamma$. The electron ($e^-$) and a positron ($e^+$) are the decay products from neutron thru double beta decay.

It is so obvious that electrons are made from photons or γ photons! It is another piece of vital evidence supporting the author's deducing that our big bang is γ ray bust!

The γ ray photons are building blocks for everything existing in our universe!

Let's review some of the interesting nuclear decay here. An example of electron emission (**β^- decay**) is the decay of carbon-14 into nitrogen-14:

$^{14}C_6 \rightarrow {}^{14}N_7 + e^- + v^-$. From the equation, it shows that the daughter nucleus contains one proton more than the parent nucleus. Thus, this type of decay moves the parent atom to daughter atom who is next higher neighbor in the periodic table. Because the equation indicates that the daughter nucleus contains one proton more than the parent nucleus! Such birth panics many people who then, create anti-particle or antineutrino to "cure" the "non-sense". The anti-neutrino seems to be the foundation for the concept of anti-matter that dominant the pop culture thereafter.

An example of the beta plus decay (**β^+**) is given by, $^{23}Mg_{12} \rightarrow {}^{23}Na_{11} + e^+ + v$

From the equation it is clear that the daughter nucleus has one proton less than the parent nucleus.

The neutrinos are similar to the electron, with one difference: neutrinos do not carry electric charge. There are 3 types of neutrinos: the "electron neutrino" which is associated with the electron and two other neutrinos are known as heavy electrons called the muon and the tau in current version of physics.

The only logical mechanism can be accepted here is that the neutrino is the intermediate particle between electron and photon! However, the author disagrees with the conception of anti-neutrino!

The interchange of proton and neutron will release electron and neutrino, which can be further broken down to γ ray photons. It is the reverse way of a big bang! The photon and neutrino carry no electric charge, but the electron carries positive or negative charge. The only mechanism that can be deduced is "polarization of photon bonding". The neutrino is neutral or "even bonding" of photons. The positive or negative charged electron is polarized or "non-evenly bonding" of photons, which is heavier or massive on one end than the other end! We need to reject the idea of "anti-matter" or "anti-particle" which is the cause of fundamental confusion when we study the basic mechanism of particle and mass!

The photon bonding/coupling has recently been accomplished in a human lab-a remarkable discovery called Rydberg blockade by coaxing photons into bonding together to form "molecules" – a state of matter by a group led by Harvard Professor of Physics Mikhail Lukin and MIT Professor of Physics Vladan Vuletic. [2] According to the conservation law of mass, the mass of new "molecules" must come from somewhere rather than from nothing, therefore the photon has mass. The biological photosynthesis is another common natural phenomenon where photons are captured and incorporated in mass. [3]

At the beginning of a big bang or spontaneous combustion, the temperature and pressure are fiercely high; few would argue that any other form of mass has a chance to exist other than radiation itself which is a big bang's photon! The author has subsequently linked gamma ray's photons to a big bang. [4]

When a light photon is not paired, it possesses pure kinetic energy/radiation power with little gravitational effect (massless paradox), e.g. a cosmic ray! If it is paired e.g. photon bonding or photosynthesis, it possesses positional energy/gravitation (mass paradox for some people). [4] The usual term of gravity means earth's gravity affects objects that are positioned on earth, but it really means that positioned objects are orbiting around earth simultaneously with or without physical contacts. Any objects may lose their gravity if they leave their orbital status, e.g. a free fall of raccoon. They will, however, regain their gravity when they resume their orbital status, e.g. a grounded raccoon. The single orbiting energy/power unifies our matter and particles, because particles possess orbital power/gravity as well when they are paired or orbing each other. They will, however, lose their gravity power if they leave orbital status (divorce) and fly out.

The author, therefore, deduces that there are two forms of fundamental force/energy existing in our universe. 1. Gravitation. 2. Electromagnetic power.

Many people are overwhelmed by the standard mode of particles. Following the path of Einstein's legend of "One" Equation theory ($E=MC^2$) for all, the author postulates an incredible simple solution for a busy bee- "2" particle model theory, which are photon and Neutrino! Don't trip, it is just like that!

In reality, the radiation and its bigger brothers/sisters e.g. electrons, neutrinos, protons, neutrons, and atoms are all we need to know in our reality of engineering for goods. Anything in between is just transient with few practical application. The photon and neutrino, however, are uniquely important. The author has deduced that photons can be used in optical computing [4-6] and dark matter engineering along with medicine and defense operating systems. [4]

THE UNIVERSE INFLATION AND FOCAL COLLAPSE

There is more evidence in space supporting our big bang theory. It is a swirl-like/light polarization imprint on light less than one-trillionth of a second old from our big bang. [7] Such light polarization becomes possible when a big bang's light encounters mass existing during a big bang. The big bang or spontaneous combustion lights encounter pre-existing mass which supports the author's one universe, endless cycle theory. [8] If our space were a void, the light's polarization simply will not happen. The author's model of universe-"one universe endless cycles" is also supported by discovery of Methuselah star, HD 140283 which is "older" than our current universe. HD 140283 appeared to be 16 billion years old, more than two billion years older than the rest of the cosmos (measured to be 13.78 ± 0.037 billion years old)! [9] It is simply not paradox; it is evidence supporting the universe's endless cycle in a realm of gravity and emulsion. [4]

The evolution of a big bang simply means that some of its greatest kinetic energy starts to be captured into its form of orbital energy/gravity during its cooling off stage, when galaxies start to be created. The destiny of a big bang is another big bang! The gravity along with power of emulsion (another form of gravitation) will eventually attract enough mass which will collapse into an ever increasingly big new pre-big bang dark matter!

The more mass is collected, the bigger a pre-big bang dark matter grows, which manifests stronger gravity-directly evidenced by our observation of the accelerated universe expanding! The accelerated speed is fueled by growing mass/increasing gravity power! Our observed universe has passed a new born stage of post big bang, passed a maturing stage of numerous galaxies forming. Our observed universe is entering a big bang or spontaneous combustion's preparing stage. The accelerated expanding is a direct evidence of stronger gravity pull - larger and larger mass/dark matters are forming! For those who use accelerating universe as a reason doubting the big bang theory, the author can assure those doubters that the opposite is true!

Furthermore, our galaxies are inflating because of star's photons. Without radiations from stars, our galaxies mass may collapse into dark matters which are indirectly supported by supernova effects. [10] The initial universe's inflation is powered by a big bang's photons; the secondary inflation which occurs in galaxies is fueled by their star's photons. The focal collapse is due to the fact that their stars run out of fuels, which subsequently age galaxies to their dwarf's stage/senior galaxy citizen, which are mostly made of dark matter along with fewer stars. It is evidenced by discovering dwarf galaxies in the current universe. [11]

Our universe is simply trillion and trillion light years big (infinite!), there are numerous focal collapses which are preparing for a next big bang. 14 billion light years is just a tiny drop of water in an infinitely gigantic size of our universe. Don't trip, it is just like that!

The Particle Evolution in the Universe

It has been estimated that dark matter may make up 84.5% of the total matter in the universe. [12] The big bang's radiation or high energy photons or gamma ray will penetrate any mass that exists during a big bang's explosion while some were polarized. For any gamma ray photons to be trapped or paired at any sub-nucleon level, those storming photons need to be "slowed down" first! Otherwise, the emulsion will not work with storming rays at the speed of light! The missing link is how lights can be slowed down; now the puzzle is finally solved! At a couple degrees above absolute temperature zero, the lights appear "slowed down" significantly where ubiquitous emulsion stands for a chance to work! [2] Our universe's background temperature is 2.7 Kelvin. The temperature may decrease even more when our region of universe is ripe for a next big bang; however, our background temperature will never reach absolute temperature zero. [4]

The big bang Nucleosynthesis starts with merging /coupling of big bang's photons in the soup of Cosmic Microwave photons (initial big bang's kinetic energy along with emulsion force)→ electron and neutrino→protons and neutrons/primordial nucleons formation (some big bang's kinetic energy transfer to positional or orbital energy known as gravity) → the "initial" or unstable nucleus's decay/radioactive decay → dark matter/neutrino with gravity formation/ stabilization→β decay →electric activities/x-rays and λ ray bursts → mass gathering due to dark mater's gravity → visible galaxy formation.

The dark matter/neutrino (gravity types) may go thru "initial formation" first, then "stabilization" with radioactive decay: (Cownan-Reines neutrino experiment) beta negative decay reacted with protons to produce neutrons and positrons which quickly combine with electrons, releasing gamma rays. [13] The significant part of mass after a big bang apparently are stuck in the most basic form of mass-dark matter/neutrino with gravity, the paired/bundled photons chilling with their mate-electrons. The halo of rich electric activities (e.g. x-ray), therefore, will be expected from any dark matter (gravity types) of substantial size! The author reasonably expects more activities to happen among electrons within or surround dark matters (gravity types), which may strike and trigger "dark matter (gravity types) signature" X ray. The author's view is supported by X-ray emission observed from the center of galaxies. [14]

The author believes the cosmic background wave and space radiation make up the rest of "missing matters" since most scientists turn their eyes away from photons' mass status.

Without any fancy language, the path of a particle in the universe is fairly simple: radiation/photon → dark matter/neutrino →visible matter/inorganic matter →organic matter → you → collapse →pre big bang dark mass/neutrino →big bang→→→you again! Don't trip, it is just like that!

The Energy Evolution in the Universe

The universe is a complete and enclosed system, because there is nothing existing outside our universe which may interfere with our universe's matter/ energy. The conservation law of mass and energy perfectly applies the universe as a unit. We can safely say that the universe's energy and mass are constant.

$$\mathbf{E}_{bigbang} = \mathbf{E}_k + \mathbf{E}_p$$

The $\mathbf{E}_k$ is the sum of kinetic energy presents in the current universe. $\mathbf{E}_p$ is the sum of positional energy presents in the current universe. [4]

When a light/photon is not coupled, it possesses pure kinetic energy/ radiation power with little gravitational effect. If it is coupled e.g. photon bonding or photosynthesis, it possesses positional energy/gravitation. [5] The destiny of a single photon/ray is to "marry" another photon/ray; the destiny of "married" photon/ray is to be separated from its partner. The endless cycle of "married" and "separation" of particles is closely mirroring the endless cycle of energy. The destiny of kinetic energy is to be transformed into positional energy; the destiny of positional energy is to be transformed into kinetic energy.

The conservation law of energy dictates that all nuclear power (strong or weak) existing after a big bang derives from electromagnetic power which is originated from a big bang's radiation-photons as well.

Instead of 4 basic forces, the author promotes 2 fundamental forces in the universe. 1. Gravitational/orbiting force. 2. Electromagnetic/radiation forces. Both are originated from "One" source-a photon! Don't trip, it is just like that!

The Role of Bonding in Physics and Spirit/Mind

The organic matters branch off from inorganic matters. The process requires light energy. The light energy could come from solar or earth itself. The most popular photosynthesis/photon bonding requires green chlorophyll pigments. In plants, these proteins are held inside organelles called chloroplasts, which are most abundant in leaf cells, while in bacteria they are embedded in the plasma membrane.

The first photosynthetic organisms probably evolved early in the evolutionary history of life and most likely used reducing agents such as hydrogen or hydrogen sulfide as sources of electrons, rather than water. [15] Cyanobacteria appeared later, and the excess oxygen they produced contributed to the oxygen catastrophe, [16] which rendered the evolution of complex life possible.

The organic matters are the mass possessing both kinetic energy and positional energy. To become alive and grow, the living beings require additional matter that possesses almost pure forms of kinetic energy -light energy. However, nothing will stay alive without the bonding process.

The bonding process is the key for the matter's evolution. The bonding is therefore a key factor for human life. Other than physics, the human mind needs a bonding factor too. Such bonding factor is simply known as a focus factor in the human mind/spirit. Such focus factors can be real or phony. In our mind or spiritual life, both will do the trick to provide a healthy focus preventing our mind from scattering. As a matter of fact, our universe's progress (endless circles) is solely operated by the laws of physics; whatever we believe will not make any difference! Don't trip, it is just like that!

RESULTS ANALYSIS

At the beginning of a big bang, the temperature and pressure are fiercely high; few would argue that any other form of matter may stand a chance to exist other than radiation itself which is a photon. The author has subsequently linked gamma ray's photons to a big bang. [4] At a couple degrees above absolute temperature zero, the lights appear "slowed down" significantly where ubiquitous emulsion stands for a chance to work! [2] The most primitive mass/dark matter (gravity type) starts to form thru photon bonding. [3]

The photon is a force, but if it is coupled/structured or stationed (e.g. In an apple), it satisfies the definition for mass. In other words, the mass itself is a force as well. The photon is the perfect example of a "2 in 1" divine package. The photon is the force/son of the majesty; the photon is the carrier/majesty, the photon is the majesty. Does it sound familiar? Without any fancy language, the path of particle universe is deduced as follows: radiation/photon → dark matter/neutrino →visible matter/inorganic matter →organic matter → you → collapse →pre big bang dark mass/neutrino →big bang→→→you again!

Following the path of Einstein's legend of "One" Equation for all, instead of the standard model of a particle; the author promotes "2" particle model which are photon and neutrino for practical application at the sub-electron and sub-nucleon level.

The universe is a complete and enclosed system, because there is nothing existing outside the universe which may interfere with the universe's matter/ energy. The conservation law of mass and energy perfectly apply the universe as a unit. We can safely say that the universe's energy and mass are constant.

$$\mathbf{E}_{bigbang} = \mathbf{E}_k + \mathbf{E}_p$$

The $\mathbf{E}_k$ is the sum of kinetic energy presents in the current universe. $\mathbf{E}_p$ is the sum of positional energy presents in the current universe. [4]

The conservation law of energy dictates that all nuclear power (strong or weak) existing after a big bang derives from electromagnetic power which is originated from a big bang's radiation-photons.

Instead of 4 basic forces, the author simplifies it for 2 fundamental forces for practical application. 1. Gravitation/orbiting energy. 2. Electromagnetic/ radiation forces. Both are originated from "one"-a photon!

DISCUSSION

The basic physics has provided us with a remarkable understanding of the mechanics behind everything on earth. Our nature provides us many facts which help us uncover the mechanism with the help of logic based in the laws of basic physics. In this paper, the author tries to reason things based on nature's known fact rather than imaginary scenery, such as riding on a train at light speed while turning flash lights on. No one would stand a chance to flash his light if he were to be accelerated by the Large Hadron Collider (LHC) to a light of speed.

In this paper, the author tries to use plain language with common sense (no twisting adjective terminology) to uncover our universe's mechanism by mapping its mass and energy evolution. The conservation law of mass and energy perfectly applies our universe as a unit. Therefore, the author reasons: 1. The conservation law of mass dictates that a photon is the "single" unified origin for all mass. 2. The conservation law of energy dictates that photon energy/radiation is the "single" unified origin for all electromagnetic power along with nuclear forces as well as photon's mass which serves the "single" unified origin for all gravitational power that are exerted by our universe's mass (detectable or undetectable). The conservation law of mass and energy perfectly supports Einstein's "One" Equation for all, because our universe's story is all about changing in mass or energy, it is just like that- incredibly simple!

If we were to set aside our urge for sophistication and fantasy for a genie, following the path of Einstein's legend of "One" Equation theory ($E=MC^2$), you will see that a photon is the "single" unified worship for all branches of science along with philosophy and religion. The universe is just another glorified "two in one package"-space (void) and photons. Time is a smart organizer invented by high level organic intelligence (not in a raccoon's universe). Our physical life is part of a photons' evolution while our spiritual mind is bouncing between reality and fairy tales.

The author promotes photon research for its practical application in future computer, medicine, and military along with the universe's mechanical study.

REFERENCES

1. Weinstein EA; Ordonez AA; DeMarco VP; Murawski AM et al. (2014). "Imaging Enterobacteriaceae infection in vivo with 18F-fluorodeoxysorbitol positron emission tomography". Science Translational Medicine 6 (259).doi:10.1126/scitranslmed.3009815. PMID 25338757.

2. M. Lukin Harvard and V. Vuletic MIT Nature Sep 25, 2013.

3. Edwin Zong, The Origin of Particles, Journal of Nuclear and Particle Physics, Vol. 5 No. 3, 2015, pp. 45-51. doi: 10.5923/j.jnpp.20150503.01.

4. Edwin Zong, 2015 The Real Universe, Journal of Nuclear and Particle Physics, Vol. 5 No. 2, 2015, pp. 21-29. doi: 10.5923/j.jnpp.20150502.01.

5. McAulay, Alastair D. (1991). Optical Computer Architectures: The Application of Optical Concepts to Next Generation Computers. New York, NY: John Wiley & Sons. ISBN 0-471-63242-2.

6. Mortaza Noshad, Amin Abbasi, Reza Ranjbar, Reza Kheradmand. (2012) Novel All-Optical Logic Gates Based on Photonic Crystal Structure. Journal of Physics: Conference Series350 (2012) 012007.

7. Gary Robbins 2014 Scientists see 'fingerprint' of Big Bang The San Diego Union-Tribune.

8. Edwin Zong. 2014 One Universe, Endless Cycles. Science Research. Vol. 2, No. 5, 2014, pp. 105-110. doi: 10.11648/j. sr.20140205.15.

9. A. J. Gallagher et al. (2010). "The barium isotopic mixture for the metal-poor subgiant star HD 140283". Astronomy and Astrophysics 523: A24. arXiv: 1008.3541. Bibcode: 2010A&A...523A..24G. doi:10.1051/0004-6361/201014970.

10. Ott, C. D. et al. (2012). "Core-Collapse Supernovae, Neutrinos, and Gravitational Waves". Nuclear Physics B: Proceedings Supplement 235: 381. arXiv: 1212.4250. Bibcode: 2013NuPhS.235..381O.doi:10.1016/j.nuclphysbps.2013.04.036.

11. Josh Simon (2005). "Dark Matter in Dwarf Galaxies: Observational Tests of the Cold Dark Matter Paradigm on Small Scales".

12. Sean Carroll, Ph.D., Cal Tech, 2007, The Teaching Company, Dark Matter, Dark Energy: The Dark Side of the Universe, Guidebook Part 2 page 46, Accessed Oct. 7, 2013.

13. C. L Cowan Jr., F. Reines, F. B. Harrison, H. W. Kruse, A. D McGuire (July 20, 1956). "Detection of the Free Neutrino: a Confirmation". Science 124 (3212): 103–4. Bibcode: 1956Sci...124.. 103C. doi:10.1126/science.124.3212.103. PMID 17796274.

14. Tauris & van den Heuvel (2006), "Formation and evolution of compact stellar X-ray sources", In: Compact stellar X-ray sources. Edited by Walter Lewin & Michiel van der Klis. Cambridge Astrophysics Series, p.623-665, DOI: 10.2277/0521826594.

15. Olson JM (May 2006). "Photosynthesis in the Archean era". Photosyn. Res. 88 (2): 109–17. doi:10.1007/s11120-006- 9040-5. PMID 16453059.

16. Buick R (August 2008). "When did oxygenic photosynthesis evolve?". Philos. Trans. R. Soc. Lond., B, Biol. Sci. 363 (1504): 2731–43. doi:10.1098/rstb.2008.0041. PMC 2606769. PMID 18468984.

Chapter 7

SINGLE PARTICLE POTENTIALS AND THREE-BODY FORCES

H. M. M. Mansour

Physics Department, Faculty of Science, Cairo University, Egypt

ABSTRACT

The single particle potentials for both asymmetric nuclear matter and pure neutron matter are presented. The Brueckner-Hartree-Fock (BHF) approximation + two body density dependent Skyrme potential which is equivalent to three- body interaction are used. Various modern nucleon-nucleon (NN) potentials are used as follows: CD-Bonn potential, Nijm1 potential, Reid 93 potential and Argonne V_{18} potential are used in the framework of the Brueckner-Hartree-Fock approximation (BHFA).

INTRODUCTION

One of the most challenging aims of nuclear physics and nuclear astrophysics is to study the equation of state (EOS) and single particle (s.p.) properties of asymmetric nuclear matter in a wide density range. At densities around and below the nuclear saturation density the properties of asymmetric nuclear matter and their isospin-asymmetry dependence are closely related to the structure, decay and collective properties of heavy nuclei and neutron-rich nuclei away from the nuclear stability line, such as the radius, the neutron skin thickness and the density distribution. The properties of asymmetric nuclear matter can be predicted by adopting various nuclear many-body approaches, including phenomenological methods and microscopic approaches. In the phenomenological methods such as the Skyrme-Hartree- Fock framework and the relativistic mean field theory, the many-body correlations in nuclear medium have been incorporated implicitly and effectively into the parameters of the adopted effective interactions. Microscopic many-body approaches start from the realistic nucleon-nucleon (NN) interactions which are determined by reproducing the experimental NN phase shifts. It is well known that the nonrelativistic microscopic approaches adopting realistic two-body NN interactions miss the empirical saturation point of nuclear matter, and three-body forces (TBF) are required. In recent years, the EOS and s.p. properties of asymmetric nuclear matter have been investigated extensively within the framework of various microscopic approaches including the Brueckner-Hartree-Fock (BHF) and the extended BHF approaches [1, 2-7], the relativistic Dirac-BHF (DBHF) theory [8-15], the in-medium T-matrix and Green function methods [16-27], and the many-body variational approach [28-33]. Up to now, several different kinds of TBF models have been adopted in nuclear microscopic many-body calculations. One is the semi phenomenological TBF such as the Urbana TBF [34]. Another TBF model adopted in the Brueckner theory is the microscopic one [35-37] based on the meson exchange theory for the NN interactions. In the BHF calculation, the TBF contribution has been included by reducing the TBF into an equivalent effective two-body interaction according to the standard and extensively adopted scheme [37]. As an important input for calculations of nuclear structures and simulations of heavy-ion reactions, the single-nucleon potential $U_{n/p}$ (k) itself can also be obtained. The results of BHF calculations depend on the choice of single particle potential U(k). The conventional choice, which assumes a single-particle potential U = 0 for single-particle states above the Fermi level, and approximate the energies by the kinetic energy only [15], and U is self-consistent BHF potential for k < kF, while the continuous choice for which U is extended to k > kF, which leads to an enhancement of correlation effects in the medium and tends to predict larger binding energies for nuclear matter and pure neutron matter than the conventional choice.

In a previous work [39], the bulk properties of cold and hot asymmetric nuclear matter were calculated in the framework of (BHF interaction + two-body density dependent Skyrme potential which is equivalent to three body force). In the present work we extend the calculation to present the single particle potentials for the proton and neutron using modern nucleon-nucleon (NN) potentials in the framework of (BHFA).

In the next section we show the model used and in section 3 the results are presented.

Single-Particle Potential (BHFA)

The G-matrix is defined by:

$$G(\omega) = V + V \frac{Q}{\omega - H_o + i\eta} G(\omega)$$

(1)

This is known as the Beth-Goldstone equation; here ω is the starting energy which is usually the sum of the single particle energies of the states of the interacting nucleons

$$\omega = e\,(k) + e\,(k') \tag{2}$$

The dashed and wiggly lines denote the bare interaction V and G matrix respectively.

V is the bare NN potential, η is infinitesimal small number, $H_{\circ}$ is the unperturbed energy of the intermediate scattering states and Q is the Pauli projection operator, it projects out states with two nucleons above the Fermi level, it is given by:

$$Q\,(k,\,k') = (1- \Theta_F\,(k))\,(1- \Theta_F\,(k')) \tag{3}$$

Where $\Theta_F\,(k) = 1$ for $k < k_F$ and zero otherwise, $\Theta_F\,(k)$ is the occupation probability of a free Fermi gas with a Fermi momentum k_F. Eq.(1) sums the ladder-type diagrams depicted in fig.(1), where the left-hand side represents the BHFA, it is the sum of the HF contribution, and all the diagrams obtained by adding an arbitrary number of interactions between particles.

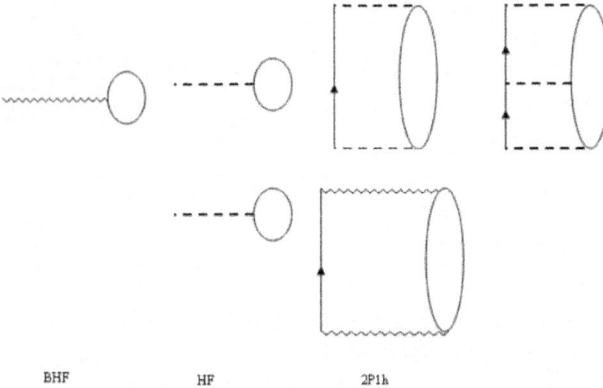

BHF   HF   2P1h

Figure 1: Graphical representation of the Bethe-Gold-stone equation

In the Brueckner-Goldstone expansion, the average binding energy per nucleon is expanded in a series of terms as the following

$$U(k)=\begin{cases} \sum_{k'<k}\langle kk'|G(e(k)+e(k'))|kk'\rangle & k<k_F \\ 0 & k>k_F \end{cases} \tag{4}$$

where |kk' > refer to antisymmetrized two-body states. This first order is known as the Brueckner-Hrtree-Fock approximation (BHFA). To completely determine the average binding energy one has to define the single particle potential U(k) which contributes to the single particle energies appearing in the G-matrix elements. The structure of the expression (4) suggests choosing the following BHF single particle potential

$$U(k) = \sum_{k' \langle k_F} \langle kk' | G(e(k)+e(k')) | kk' \rangle$$

(5)

$$\frac{E(k)}{A} = \sum_{k \langle k_F} \left\{ \frac{k^2}{2m} + \frac{1}{2} U(k) \right\}$$

$$= \frac{4}{\rho} \frac{1}{2} \int_0^{k_F} \frac{4\pi k^2}{(2\pi)^3} \frac{k^2}{2m} (\frac{k^2}{2m} + e(k))$$

$$= \frac{3k_F^2}{10} + \frac{3}{2k_F^3} \int_0^{k_F} k^2 dk U(k)$$

(6)

The G-matrix itself depends on U (k) through the starting energy ω, defined in eq.(2) and the lowest order approximation (4) along with choice (5) for the single particle potential is often known as the lowest order Brueckner theory.

The single particle energy e (k) is defined as

$$e(k) = K.E + U(k) = \frac{\hbar^2 k^2}{2m} + U(k)$$

(7)

where K. E is the kinetic energy. The conventional choice for the single particle potential has been to take the BHF potential (eq. (5)) for hole states ($k < k_F$) and zero for particle states ($k > k_F$).

$$U(k) = \begin{cases} \sum_{k' \langle k} \langle kk' | G(e(k)+e(k')) | kk' \rangle & k \langle k_F \\ 0 & k \rangle k_F \end{cases}$$

(8)

Thus introducing a quite large discontinuity in the single particle spectrum at the Fermi surface. However, due to the unphysical discontinuity at the Fermi surface this auxiliary potential cannot be directly related to the average potential felt by a particle or a hole. Moreover, many other interesting properties can be derived such as the momentum distribution and the effective mass which is properly described using a continuous spectrum across the Fermi surface. This was the main motivation which led [38] to the introduction of the continuous choice for the single particle potential thus treating particles and holes in a symmetrical way. The use of the continuous choice potential implies that the G-matrix elements needed in the self-consistent calculation are complex and the prescription advocated is

$$U(k) = \mathrm{Re} \sum_{k' \langle k_F} \langle kk' | G(e(k) + e(k')) | kk' \rangle$$

(9)

Eqs. (1) and (7) represents the main equations that we want to solve self-consistently. In order to obtain such a self-consistent solution one often assumes a quadratic dependence of the single-particle energy on the momentum of the nucleon in the form

$$e(k) = \begin{cases} \dfrac{\hbar^2 k^2}{2m^*} + \Delta & k \le k_F \\ \dfrac{\hbar^2 k^2}{2m^*} & k \rangle k_F \end{cases}$$

(10)

where m^* is the effective mass of the nucleon and Δ is a constant. Starting with an appropriate choice for the parameters for the effective m^* and the constant Δ, one can solve the Bethe-Goldstone equation and evaluate the single-particle energy. The parameters m^* and the constant Δ can then be readjusted in such away that the parameterization eq. (10) reproduces these two energies. This procedure is then iterated until a self-consistent solution is obtained. The parameterization of eq. (10), however, is useful not only to simplify the self-consistent solution of the BHF equations; it also leads to a simplification of the numerical solution of the Bethe-Goldstone equation.

SINGLE-PARTICLE POTENTIAL (TBF)

In the present work one may introduce a Skyrme effective interaction density dependent term in addition to the BHF single particle potential in the previous section [39].

$$V(\bar{r}_1, \bar{r}_2) = \sum_{i=1}^{4} t_i (1 + x_i P_\sigma) \rho^{\alpha_i} \delta(\bar{r}_1 - \bar{r}_2)$$

(11)

This is a two-body density dependent potential which is equivalent to three body interaction. Where t_i and x_i are interaction parameters, P_σ is the spin exchange operator, ρ is the density, r_1 and r_2 are the position vectors of the particle (1) and particle (2) respectively. $\alpha_i = (1/3, 2/3, 1/2$ and $1)$. Using the above potential for asymmetric nuclear matter the following correction for the single particle potential U is used:

$$\Delta U = (3/4)\, t_1\, \rho^{4/3} + (3/4)\, t_2\, \rho^{5/3} \tag{12}$$

And for pure neutron matter we obtain the following correction

$$\Delta U = (1/2)\, t_1 \rho^{4/3}\, [1 - x_1] + (1/2)\, t_2\, \rho^{5/3}\, [1 - x_2] \tag{13}$$

Results

After adding TBF corrections to the BHF calculations it was found that among these corrections the best results to obtain the empirical saturation point when $\alpha_i = (1/3$ and $2/3)$. The parameters are the same as given in Ref. [39]. Figure (1) shows a comparison between the EOS in the framework of the BHFA in (A) and EOS in the framework of BHFA+ two body density dependent Skyrme interaction in (B) by using the CD-Bonn potential [40] and Reid93 potential [41] for conventional Choice. This shows an adjustment of the saturation point location and value by adding the TBF.

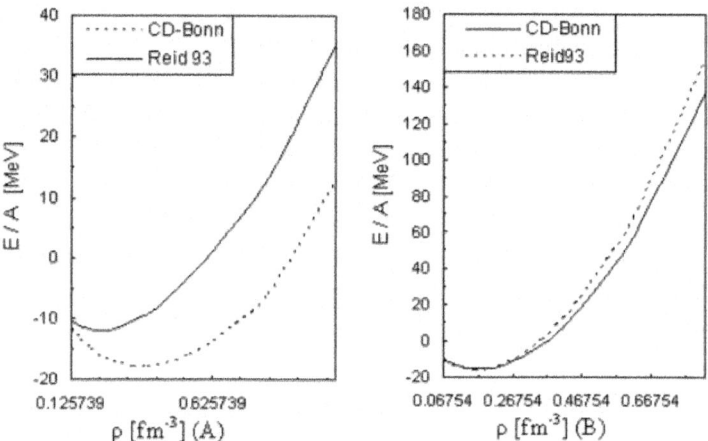

Figure 2: The EOS in the framework of the BHFA in (A) and EOS in the framework of BHFA+ two body density dependent Skyrme interaction (which is equivalent to three body force) in (B) by using the CD-Bonn potential and Reid93 potential for conventional Choice

The single particle potential of e.g. a proton U is defined together with kinetic energy as the energy required to remove this proton from the nuclear system leaving a hole in the state. The results of BHF calculations depend on the choice of the single particle potential U (k) in the "standard or conventional" choice, U = 0 for k > k_F and U is the self-consistent BHF potential for k < k_F, the alternative "continuous" for which U is again the self-consistent BHF potential. The single particle potential for asymmetric nuclear matter is calculated using the CD-Bonn potential, the Argonnev$_{18}$ [42], the Nijm1 [41] potential and the Reid 93 potential for conventional and continuous choices. The dependence of the single particle potential on the momentum k for asymme-tric nuclear matter at k$_F$ = 1.333 fm^{-1} for various potentials for conventional and continuous choices are shown in figure (2). The results for all potentials are approximately similar in the conventional choice as well as the continuous choice. The single particle potential increases with increasing the momentum k (U is directly proportional to k).

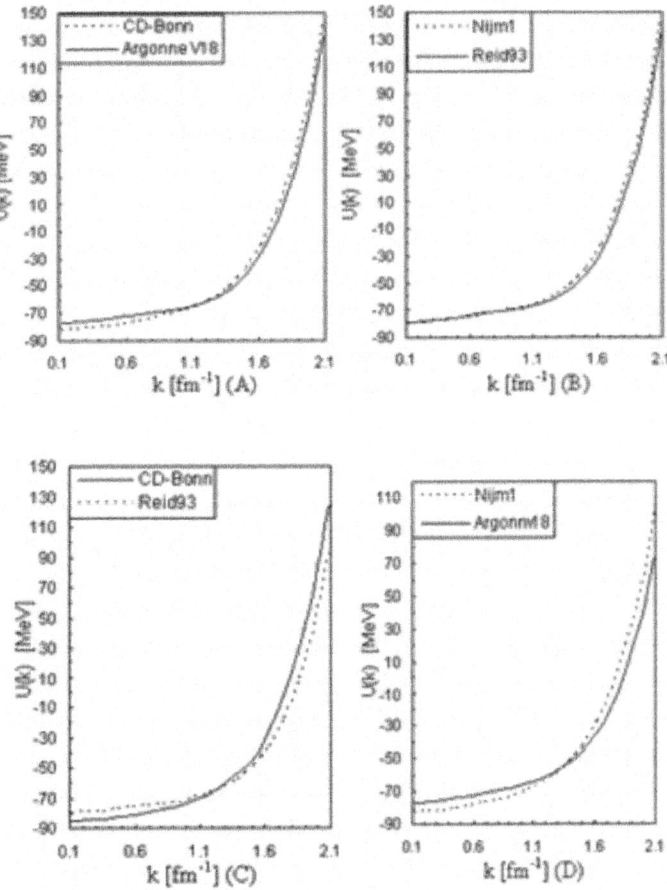

Figure 3: The single particle potential for asymmetric nuclear as a function of momen-tum k at (k$_F$ = 1.333 $^{fm-1}$) for different potentials (A and B) for conventional choice, (C and D) for continuous choice

In figure (3.3) the dependence of the single particle potential on the momentum k for the pure neutron matter at $k_F = 1.333$ fm⁻¹ is plotted for various potentials for conventional choice the CD-Bonn potential (A), the Nijm1 potential (B), the Reid 93 potential(C) and the Argonne v_{18} potential (D). In figure (3.4) we get the same as above but for continuous choice. It is observed that the results for all potentials are approximately similar in the conventional choice and the continuous choice the results for different potentials have the same behavior. The single particle potential increases with increasing the momentum k. In conclusion we claim that a simple and analytic term for the three –body potential has corrected the deficiency in the BHF calculation in the sense that we get the right value for the energy at the saturation point, the symmetry energy increases with increasing the density and good agreement with other theoretical works for the physical quantities of relevance to the nuclear and neutron matter as shown in previous works [39, 40-46]. For completeness we present here the calculation for the single-particle potential.

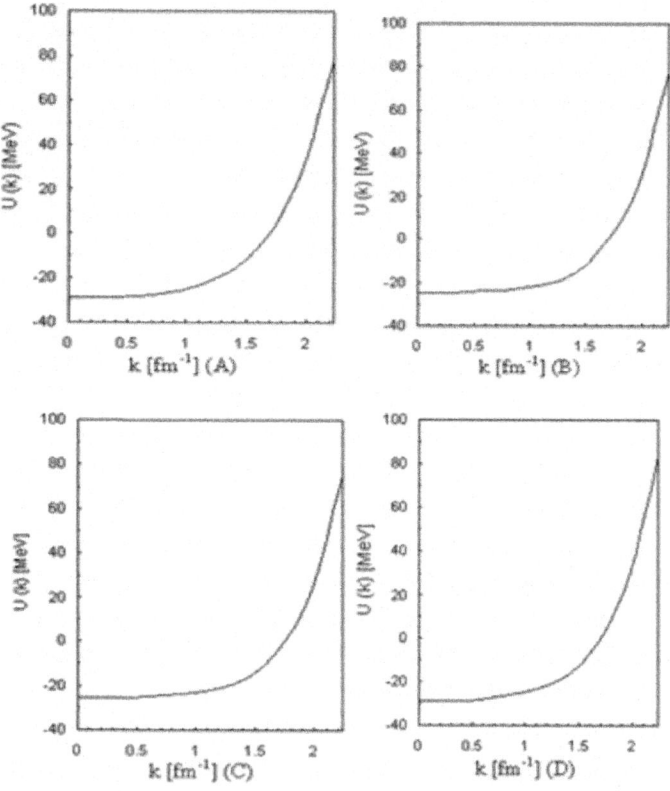

Figure 4: The single particle potential for the pure neutron matter in [Me V] as a function of momentum k at ($k_F = 1.333$ fm⁻¹) for continuous choice of different potentials

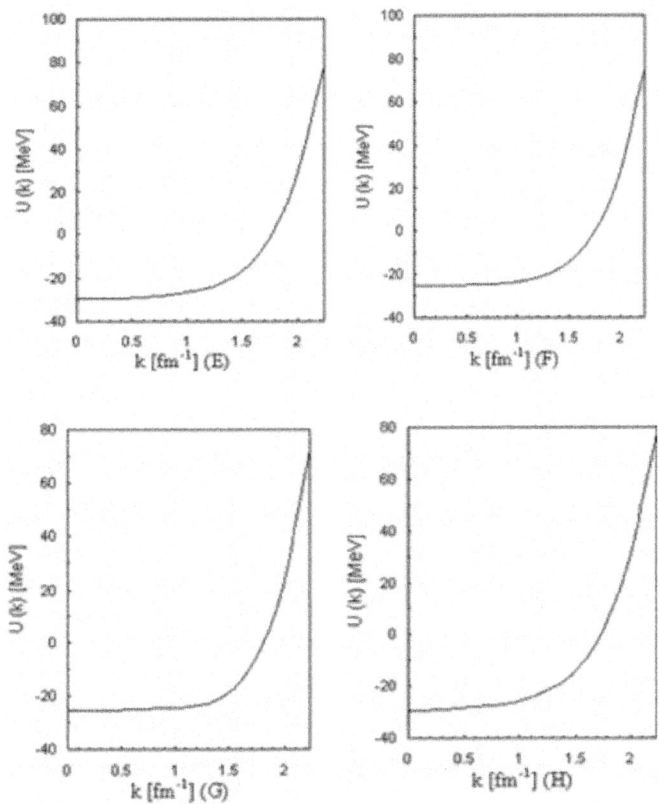

Figure 5: The single particle potential for the pure neutron matter in [Me V] as a function of momentum k at ($k_F = 1.333$ fm⁻¹) for continuous choice of different potentials

REFERENCES

1.   I. Bombaci, U. Lombardo, Phys. Rev. C 44, 1892 (1991).

2.   W. Zuo, I. Bombaci, U. Lombardo, Phys. Rev. C 60,024605 (1999).

3.   W. Zuo, A. Lejeune, U. Lombardo, J.F. Mathiot, Eur.Phys. J. A 14, 469 (2002).

4.   W. Zuo, L.G. Cao, B.A. Li, U. Lombardo, C.W. Shen, Phys. Rev. C 72, 014005 (2005).

5.   W. Zuo, U. Lombardo, H.J. Schulze, Z.H. Li, Phys. Rev.C 74, 014317 (2006).

6.   I. Vida~na, C. Provid^encia, A. Polls, A. Rios, Phys. Rev.C 80, 045806 (2009).

7.   I. Vida~na, A. Polls, C. Provid^encia, Phys. Rev. C 84,062801 (2011).

8. H. Huber, F. Weber, M.K. Wiegel, Phys. Rev. C 51, 1790 (1995).
9. C.H. Lee, T.T.S. Kuo, G.Q. Li, G.E. Brown, Phys. Rev.C 57, 3488 (1998).
10. E.N.E. van Dalen, C. Fuchs, A. Faessler, Nucl. Phys. A744, 227 (2004).
11. E.N.E. van Dalen, C. Fuchs, A. Faessler, Phys. Rev. C 72, 065803 (2005).
12. E.N.E. van Dalen, C. Fuchs, A. Faessler, Phys. Rev. Lett.95, 022302 (2005).
13. Z.Y. Ma, J. Rong, B.Q. Chen et al., Phys. Lett. B 604,170 (2004).
14. P. Krastev, F. Sammarruca, Phys. Rev. C 73, 014001 (2006).
15. T. Klahn, D. Blaschke, S. Typel et al., Phys. Rev. C 74,035802 (2006).
16. T. Frick, H. M¨uther, A. Rios, A. Polls, A. Ramos, Phys.Rev. C 71, 014313 (2005).
17. Kh. Gad, Kh. S.A. Hassaneen, Nucl. Phys. A 793, 67(2007).
18. A. Rios, A. Polls, W.H. Dickhoff, Phys. Rev. C 79, 064308(2009).
19. A. Rios, V. Soma, Phys. Rev. Lett. 108, 012501 (2012).
20. P. Bozek, P. Czerski, Eur. Phys. J. A 11, 271 (2001).
21. P. Bozek, Eur. Phys. J. A 15, 325 (2002).
22. P. Bozek, Phys. Rev. C 65, 054306 (2002).
23. V. Soma, P. Bozek, Phys. Rev. C 78, 054003 (2008).
24. V. Soma, P. Bozek, Phys. Rev. C 80, 025803 (2009).
25. Y. Dewulf, D. Van Neck, M. Waroquier, Phys. Lett. B510, 89 (2001).
26. Y. Dewulf, D. Van Neck, M. Waroquier, Phys. Rev. C 65,054316 (2002).
27. Y. Dewulf, W.H. Dickhoff, D. Van Neck, E.R. Stoddard, M. Waroquier, Phys. Rev. Lett. 90, 152501 (2003).
28. R.B. Wiringa, V. Fiks, A. Fabrocini, Phys. Rev. C 38,1010 (1988).
29. A. Akmal, V.R. Pandharipande, D.G. Ravenhall, Phys. Rev. C 58, 1804 (1998).
30. G.H. Bordbar, M. Modarres, Phys. Rev. C 57, 714 (1998).
31. M. Modarres, G.H. Bordbar, Phys. Rev. C 58, 2781(1998).
32. G.H. Bordbar, M. Bigdeli, Phys. Rev. C 75, 045804(2007).
33. G.H. Bordbar, M. Bigdeli, Phys. Rev. C 77, 015805(2008).
34. J. Carlson, V.R. Pandharipande, R.B. Wiringa, Nucl. Phys. A 401, 59 (1983).
35. P. Grang´e, A. Lejeune, M. Martzolff, J.F. Mathiot, Phys. Rev. C 40, 1040 (1989).
36. W. Zuo, A. Lejeune, U. Lombardo, J.F. Mathiot, Nucl. Phys. A 706, 418

(2002).

37. Z.H. Li, U. Lombardo, H.J. Schulze, W. Zuo, Phys. RevC 77, 034316 (2008).

38. Maclleidt R. Adv. Nucl. Phys., 19, 189(1989).

39. H.M.M. Mansour and A. Gamoudi, Physics of atomic nuclei 75,430(2012).

40. Machleidt R. Phys. Rev. C63 (2001), 024001.

41. Stoks V. G. J, Klomp R. A. M, Terheggen C. P. F and de Swart J. J, Phys. Rev. C49 (1994), 2950.

42. Wiringa R. B, Stoks V. G. J. and Schiavilla R, Phys. Rev C51 (1995), 38.

43. Hesham M.M. Mansour, Khaled S.A. Hassaneen, Journal of Nuclear and Particle Physics 2(2) (2012) 14.

44. Khaled Hassaneen and Hesham Mansour, Journal of modern physics 4(5B),37(2013).

45. Khaled Hassaneen and Hesham Mansour, Journal of Nuclear and Particle Physics 2013, 3(4): 77-96.

46. H. M. M. Mansour and Kh. S. A. Hassaneen, Physics of atomic nuclei, 2014, vol 77, No. 3, pp. 290–298.

Chapter 8

THE REAL UNIVERSE

Edwin Zong

Medicine Department, Oasis Medical Group Inc., Bakersfield, California, USA

ABSTRACT

The forces behind accelerated universe expanding have drawn many attentions among curious eyes. Some people may love to refer it as "dark" energy, the author, however, calls it "emulsion" power. The findings of strong gravitational lensing by Sgr A* indicate gravitation power/energy exist in the center of our milky way, though it is invisible. The presence of energy equal presence of mass ($M = E/C^2$, C^2 is constant). However, current perception of particles can be massless e.g. the photon [1] which create a centerpiece of confusion when we come to study dark matter and origin of the universe. In this paper, the author will explain why all particles possess mass which is supported partially by 2013 Nobel Prize in Physics- the Higgs mechanism, which gives mass to fundamental particles. The big bang is the greatest event happened in the universe; it is unthinkable that such event left no trace in space; we must be blindfolded by something. The author will point out what to look for so that we won't miss out on the Universe greatest show. Few will argue our matters (whether visible or invisible) are all made from particles. The positional energy or orbiting energy/gravitation power unifies our matter and particles, because particles possess orbiting power as well when they are coupled. The mystifying gravitation as some sort of supernatural phenomenon (e.g. space-time or gravitational particle) misleads our entire generation since Newton expired. The author also will explain the reasons that dark matter/ black hole is made from gamma ray nucleosynthesis or photosynthesis γ type! The author will explain the reasons that our current background cosmic ray/ gamma ray originated directly from a big bang (for most part) as well! The Newtonian Physics will unite all science branches. The particle physics will lead artificial intelligence to its terminal end in the universe civilization of all kind! The primary objective of this study is to uncover the real mechanic of dark matter, big bang and evolution process of the universe in the discipline of basic physics.

INTRODUCTION

All matters (include fundamental particles) intrinsically possess positional energy/power and kinetic energy/power. For better understanding, the author here describes all positional energy as orbiting energy. The usual term of gravity means earth's gravity affects objects that are positioned on earth, but it really means that positioned objects are orbiting around earth simultaneously with or without physical contacts. The single orbiting energy/power unifies matter and particles, because particles possess orbiting power/gravity as well when they are coupled. The mystifying gravitation as some sort of supernatural phenomenon (e.g. space-time or gravitational particle) misleads our entire generation since Newton was born. The author can assure you that those orbiting energy/power of particles follow similar laws that apply their heavier counterparts (e.g. an apple's gravity power). The nature is never stingy in providing facts for us! Here is a very common phenomenon in an apple farm, the un-coupled light photon is captured and coupled in an apple tree formation. Such light photon now becomes part of an apple tree weight. The gravitation of an apple tree is so obvious, but it doesn't deny the miniscule gravitation contribution of a single light photon. When a light photon is not coupled, it possesses pure kinetic energy. If it is coupled, it possesses positional energy/gravitation, though it loses its freedom as an uncoupled photon. In a sense, the nature provides us a perfect example of dark matter formation in our own apple farm! The striking uncoupled photon disappears right in front of our eyes when it is coupled in apple tree. It serves a terrific example that a visible matter (free solar light) turns into invisible matter (part of mass possessing gravity). The photosynthesis begins when photon is absorbed by proteins called reaction centres that contain green chlorophyll pigments. In an apple tree, these proteins are held inside organelles called chloroplasts, which are most abundant in its leaf cells. The chemical reaction of photosynthesis: CO_2+H_2O+Solar photons$\rightarrow C_6H_{12}O_6$ (glucose)+O_2.

The apple leaf cells "darken" the solar photon by chemical reaction that turns a pure kinetic energy (photon) into a pure positional energy (photon coupling). The slightly weight gaining or "added" gravitation energy/power is resulted from pure kinetic "un-coupled" photons that are captured and "darkened" by chloroplasts. The "coupled" photons simply mean "stable" particles which are "settled". The "stable" mass can orbit another "stable" mass which projects gravity effects of most significance! For instance, if mass "itself" is not stable (e.g. exploding firework, which will not orbit around any mass till its kinetic energy settles into positional energy, then it will fall and orbit with earth-recapturing gravity effects!). It is the same reason that dark matters are stable matters or "settled" particles of some sort, which casts gravity effects. Similarly, the solar "uncoupled" photons must be "coupled or settled" first before it can be part of gravity effects of an apple tree (not limit to the one Newton once sitting under!). The "dark matter of earth type" and its formation is no stranger to an earthling if we know where to have our eyes on. As a matter of fact, the different character of apple leave and dark matter is "small". Without outside lighting, the apple leaf without active metabolic activity becomes "invisible" though it has absorbed visible solar lights/solar light coupled, so is dark matter. With outside lighting, the apple leaf becomes visible because it reflexes certain lights out, dark matter simply absorbs those lights, so dark matter remains invisible. If mass is not "coupled or settled", it will be "influenced" or "bent" by nearby mass (whether stable or not)-gravity effects of less significance.

Uncoupled particles/flying ray in a straight direction will not be able to cast significant gravity effects on other objects! Thus, uncoupled photos cannot be part of gravity of any sort, other than catching attention from a pair of curious eyes! Therefore, the apple tree contains many of those coupled photons, not uncoupled photons! The apple tree is a perfect example of objects that possess positional energy/gravitation power (such example does not limit to the one fell on young Newton's head), but it also has its root in pure kinetic energy!-uncoupled light photons which possess no gravitation power. There is no space-time or mysterious particle between uncoupled light photons and coupled photons or an apple and newton's head!

Those orbiting energy of particles are same as earth gravity power, which should be scaled in gravity term.

Furthermore, the gravity affects particles in same way as it does to their heavier mass brothers, even though the gravity effects on heavier matter are sometimes negligible (e.g. a flamingo's gravity effects towards to our earth, but it is not vice versa). It's that reason, any study of particle must be in a lab environment of 0G0K (Zero Gravity Zero Kelvin). The study of particles' behavior in a non-0G0K environment is invalid [2] [3]. 0G0K is beyond human reach, but it will not stop us from understanding those little creatures. The author can assure you, when they are coupled, all particles orbit in a circular way, as it appears for any orbiting celestial body. When particles are liberated from their orbits, they will fly out in a straight way in a void space. The flying path of liberated particle can be bent. The particle will be affected by gravity if foreign particles or mass exist, exemplified as gravitational lens phenomenon, that refers to a distribution of matter (such as a cluster of galaxies) between a distant source and an observer, that is capable of bending the light from the source, as it travels towards the observer. That is one of the direct evidence that light particle possesses mass. The gravity only exists among masses whether they are big or negligible small. The denial of gravity effects as part of positional energy/power is not worth for a scientific debate. Because if it were true, the rocket scientists would be modern tailors for emperor's new cloth (Emperor's New Clothes by Hans Christian Andersen).

The light particles can be absorbed as well; it can be absorbed by inorganic and organic material. The liberated particles advance in a wave way, a straight direction in a void space. We often describe them as electromagnetic radiation or ray. We often know maters better, because many of them are visible to our naked eyes, e.g. soccer. When matter breaks down to an invisible level, many panic. Some create principle of uncertainty for those invisible devils, but the author assures you, invisible particles follow same laws of physics and mathematics odds of random. Many bizarre behaviors of observed particles are resulted from contaminated labs (non-zero gravity, non-absolute zero temperature).

The conservative of energy dictates that liberated particles possess greater kinetic energy when they leave their orbit. The orbiting energy they once possess now fuels their kinetic energy. The destiny of liberated particles is to reunite other particles. The reunion will eventually lead to recapture its positional energy/power or gravity! The destiny of coupled "stable" particles is to be separated in a bang of some sort. The fate of particle closely mirrors our universe visible matters. The destiny of cosmos objects is destruction/big bang, every destruction/big bang breeds new born of cosmos objects/construction. The universe is in an endless cycle of matters/energy, where universe law applies both visible and invisible matters [2] [3].

There is no mystery in the universe! Other than gravity profiling, we also need to know the intrinsic temperature of particle. Particle physics are the cutting edge of all science, but its focus should be on profiling particles in a correct way-quantifying its G and K.

BIG BANG, Γ-RAY, DARK MATTER, BLACK HOLE

The universe is a complete and enclosed system, because there is nothing existing outside universe which may interferes universe matters/energy. The conservation of mass and energy perfectly apply universe as a unit. We can safely say that universe energy and mass are constant. The universe is never a boring place, there are many events happening in the universe, but nothing is more significant than a big bang. First, few will argue the big bang gives out the greatest kinetic energy. The question is where the big blast of kinetic energy goes today? The kinetic energy cannot disappear without trace. Second, the big bang spits out matters which carry tremendous energy, where are those high energy matters today? Again, the matters cannot disappear without a trace.

Admittedly, much of a big bang's kinetic energy transforms into positional energy (conservation of energy) after the big bang. This is directly evidenced by numerous stars, planets etc. All those positional energy possessed by today's celestial mass originated from the big bang's kinetic energy; however, it is not a complete story here. There are numerous circulating celestial mass which carry tremendous kinetic energy besides their positional energy. Furthermore, there come cosmic rays! Those cosmic rays scatter with no pairing; they are free standing loners. The cosmic rays carry little positional energy but tremendous kinetic energy. Among them, the Gamma Ray is most significant kinetic energy form existing in the universe. The big bang Energy can be safely construed as pure kinetic energy (those kinetic energy is transformed from positional energy possessed in pre-big bang Mass). Now we can put all the pieces of puzzle together, it solves a very simple question in a simple equation.

$$\mathbf{E}_{bigbang} = \mathbf{E}_k + \mathbf{E}_p$$

The $\mathbf{E}_k$ is the sum of kinetic energy presents in the current universe. $\mathbf{E}_p$ is the sum of positional energy presents in the current universe.

The matter that carries kinetic energy can be considered "visible", because we have a way to detect them directly, e.g. its body temperature. The temperature is a good example of kinetic energy. The other good example of kinetic energy is orbiting planets. The speed prevents them from falling into their orbiting mass. Many objects possess kinetic energy also possess significant positional energy e.g. sun. The sun carries kinetic energy and it has gravity (a form of positional energy). Many objects possess kinetic energy, but own little or none positional energy. Luckily, they possess kinetic energy, so we can detect them. Those creatures perfectly meet the needs of the big bang explosion. Among all those creatures, we can clearly see nothing is more qualified than Liberated Gamma Ray/Particles to fit into this role. The liberated Gamma Ray/Particles stand for free from orbiting (e.g. Radioactive decay).

What confuse many scientists today are objects that possess positional energy but little kinetic energy. Some of those objects that possess positional energy also possess kinetic energy e.g. sun. Those are considered visible and detectable. What about objects possessing positional energy but have little or no kinetic energy? Furthermore, when their positional energy powerful enough, they will absorb kinetic energy of any kind, pure or partial. They will have no lights reflex at all. If those objects present right in front of us, we can touch them but not able to see them. While we touch those objects, we will most likely be absorbed into them and dissembled to particles. If they present far away, beyond our physical reach, the objects could be a nightmare for scientists. The author agrees that dark matter is a perfect name for them. They are matters but invisible.

The big explosion/big bang is the most extreme event happening here and there in the universe. It experiences moment of silent/quiet followed by a gigantic explosion (positional energy transforms into kinetic energy in energy term for a blast). Nothing is more qualified than dark matter to be a pre-bang mass which meets the requirement for feeding a big bang.

The black hole is made of dark matters. The so-called dark/black simply means invisible to human eyes. Invisible doesn't equal to non-existing. Dark matters emit little lights, their temperature are very low as they are evidenced by their presence in the center of Milky Way. A black hole of 4.5×1022 kg (about the mass of the Moon) would be in equilibrium at 2.7 kelvin, absorbing as much radiation as it emits [4]. For comparison, at the core of our sun, the temperature can reach more than 27 million degrees F (15 million degrees C). The sun only holds our solar system together, but the black hole holds our milky way together. Our milky way is said to have 100 billion stars. It is undoubtedly to say, our black hole possesses enormous amount of gravitation power, which means enormous amounts of mass. The black hole is no mystery. The gravitation power is part of positional power. However, the recent observations of the star S14 (S0-16) circulates the center of our milky way indicate that our black hole radius is no more than 6.25 light-hours. The only widely hypothesized type of object that can contain 4.1 million solar masses in a volume of that small size is a black hole [5].

The black hole of Milky Way is incredible cold comparing to our solar center [4]. So, given such low kinetic energy/low temperature in the black hole of Milky Way, which form does its energy exist? According to conservation of energy, it must be in its positional energy. The author can therefore assure you that dark matters/black hole possesses enormous amounts of orbiting energy/ positional energy, which is directly evidenced by Milky Way staying together. Without black hole, any galaxy will scatter away. The prediction of existence of dark matter/black hole is not genius; it is rather a common sense.

The life span of galaxy is lot longer than our solar system, which dictates our milky way's black hole cannot emit too much energy/mass out too fast or outpace its absorbing new mass/energy, while it needs to be continued to be fed by outside masses/energy. The gigantic level of black hole gravity ensures its longer life in two ways. One, few lights escape its gravity grip.

Two, tremendous amount of celestial mass are continuously pulled in. Our sun, on other hand, it sheds tremendous amounts of energy/mass continuously, but it is not fed by outside masses at any significant level. The sun has much shorter life span vs. black hole. The mass's make up of sun is different from black hole as well.

Dark matters and black holes are no magic vs. their visible cousins. The denser element means heavier element. What is heavier? It means greater gravitation power. Few will argue dark matters are compactly packed creatures. The real question is where they get such striking energy/gravitation power from? According to conservation of energy, energy cannot be created or destroyed. Their gravitation energy must come from their kinetic energy they once owned.

Who owns highest kinetic energy in the universe? The liberated particles own highest kinetic energy directly from a big bang! Now we can put final piece of puzzle in its place. Without seeing its happening, the author can assure you two things. 1. The dark matter/black hole is compacted (non-liberated or orbit-binding) gamma ray/particles nucleosynthesis! 2. The dark matter/black hole/non-liberated Gamma Ray are those original Gamma Ray/Particles directly come from big bang! The compacted or non-liberated gamma ray particles exist in dark matter of black hole is understandable, because tremendous numbers of γ-ray/particles are gathering in various dense areas (space is not vacuumed during pre or post big bang. The free standing γ-ray will be coupled or orbited among each other where they gather. The mass of orbit-binding gamma ray acquired tremendous positional power from such gathering. Those density induced γ-ray gathering/concentration triggers massive gamma ray/particles nucleosynthesis! Which is termed as dark matters/black hole. The free standing or non-orbiting γ-ray will continue to fly in a straight line (almost pure kinetic energy with little positional energy) until it meets its mate or potential coupling partner. As a matter of fact, the free standing loner γ-ray still exists in a tremendous amount in today's space- it is called cosmic ray!

Furthermore, our current cosmos Gamma ray (not all of them) is indeed from a big bang! It is partially supported by fact that the energies of gamma rays from astronomical sources range to over 10 TeV. That energy is far too large to result from radioactive decay. [6] some cosmos gamma ray are from pulsar and black hole though. However, gamma-ray pulsars and rare occurrence of gamma ray burst from black hole are very rare events in the universe. It is evidenced by recent observation: there have been only about one hundred gamma-ray pulsars identified out of about 1800 known pulsars [7] [8]. Base on observation, the sources of most GRBs are billions of light years away from Earth, implying that the explosions are both extremely energetic (a typical burst releases as much energy in a few seconds as the Sun will in its entire 10-billion-year lifetime) and extremely rare (a few per galaxy per million years) [9]. Given the facts that the universe space is infinite big and it is full of diffuse gamma radiation, it can be reasoned that the pulsars and black holes are not sufficient source for such extensive universe background gamma radiation. Secondly, the radioactive decay is not a right candidate power house for cosmos γ-ray background either. The only option left here is a big bang. The big bang provides most of those cosmos γ-ray today. The remaining cosmos γ-ray is provided by big bang indirectly. The issue remaining here is that some of those cosmos γ-ray might not come from the big bang which created our galaxy, some of those cosmos γ-ray may come from foreign big bangs if it can stay single/free long enough and it can escape being captured by cosmos mass/energy along its way.

The minute part of cosmos γ-ray may even come from alien artificial intelligence.

The author here can assure you black matter/black hole are made from gamma ray nucleosynthesis that are originated directly from big bang. The gigantic gravitation power which black hole possesses today directly comes from a big bang! There will be no other appropriate candidate source powerful enough to fuel black hole such energy at that astonishing level. It is evidenced by galaxy's black hole. In the universe space, there is no single isolated structure that is bigger than a galaxy. All galaxies are hold together by its black hole. All galaxies descend from a big bang directly, so does its black hole. As a matter of fact, all galaxies are made from gamma ray of some kind as well.

Secondly, gamma ray nucleosynthesis existing in black matter/black hole remain domicile which presents little kinetic power, however, the continue feeding of new celestial mass may trigger some of its gamma ray particles to be liberated from their orbiting position and fly out with tremendous kinetic power, which is evidenced by gamma ray burst from a black hole. It is supported by NASA SWIFT project's findings of gamma ray burst (GRB) that is directly related to a black hole.

Dark Matter, Dark Energy and Big Bang

The dark Matter is Matter that possesses little kinetic energy/power, so that it is hard to be detected by conventional detectors, e.g. telescope. The dark energy is positional energy, not kinetic energy. The kinetic energy is readily detected by our conventional detectors. However, there is a consensus that our universe dark matter makes up about 27% and 68% of the Universe is dark energy. What 68% tells us? It tells us that less than half of energy is in objects possess detectable kinetic energy. 27% mass is in Gamma Ray nucleosynthesis of some kind! The dark is no mystery. The author here will crack the nut for you. 68% of the universe energy is dark energy; it is so significant sources of energy that cannot be supplied by any other sources but a big bang itself! The greatest kinetic energy deprived from a big blast turned into positional energy shortly after explosion. The photosynthesis is a well-known nature phenomenon on earth; it is, however, a slow process. The perception of creating dark matter by two photons collision maybe a futile effort but it is worth to try. The creation and growing of dark matter is reasonably believed to be a cosmos-style slow process, otherwise, we will have a big bang every other year. The pre-big bang mass is made largely of dark matters; we will not be able to create a mini big bang without crating dark matter first in our lab.

The dark matter originated from pure kinetic energy-big bang. The process of creating dark matter is another example of nature's photosynthesis. Dark matter, similar to that apple tree, stands there with gravitation power, but it has its root in pure kinetic energy-uncoupled "light photons" which possess no gravitation power. However, the "light photons" here is not solar light, because dark matter is not an apple!

The author's point is supported by Big Bang nucleosynthesis- The first nuclei were formed about three minutes after the Big Bang! Nucleosynthesis is the process that creates new atomic nuclei from pre-existing nucleons, primarily protons and neutrons. The newly formed new atomic nuclei are perfect example of pure kinetic energy captured into positional energy/orbiting energy! The Big Bang Nucleosynthesis not only provides food for baby galaxies' visible matters but also breeds dark matters that hold galaxies visible matter together!

The positional energy possessed by dark matters apparently is necessary for galaxies to form and stay together; otherwise, all matters/energy will scatter all over places without having a chance to grow into anything significant in its size. Therefore, the "visible matters" which have chances to form must "follow" dark matters formation which occurs first, not vice versa. The author's view is supported by the dark ages of the universe — an era of darkness that existed before the first stars and galaxies ever appear. Many consider this dark age of the universe is mystery, the author think it is the most rational process in nature's evolution development.

The dark Matter forms first immediately after a big bang. The greatest kinetic energy starts to turn into positional energy during this period of time/ nucleosynthesis. Once positional energy establishes, matters starts to gather and circulate around dark matter. The dark matter merges as well, the greater dark matter, the greater mass can be collected around the dark matter. This is the model of baby galaxy's birth process.

As the author described earlier, the big bang does not occur in a vacuum space. The existing cosmos mass/cosmos ray will interact with a big bang's ray/Gamma ray, the interaction facilitate "post bang" kinetic energy transform into positional energy. The "so called" dark matter possesses so little kinetic energy for us to detect, the dark matter must be something unusual to our earthlings. Most materials which we are grown up with are matters that possess both kinetic energy and positional energy, though pure kinetic energy is very common to us, e.g. solar lights. In a macro view, the slow merging of galaxies gives out the "secret" about mass of dark matter. Because all galaxies are made of visible objects and dark matters. The gravity power of galaxies mass makes their merging possible. It is evidenced by our milky way and its "imminent" future collision partner- Canis Major Dwarf Galaxy.

The author applaud international collaboration for the Dark Energy Survey, if basic physics works for deep space, the map should demonstrate the "greater and greater" concentration of "matters or dark matters" as mass evolves or timing by humans, a phenomenon is similar to "oil water emulsion". The universe is never a place of completely "void". It is filled up with matters of two kinds – "uncoupled" and "coupled" mass at different concentration throughout space. The "uncoupled" mass e.g. cosmic ray corresponding to mostly "kinetic energy"; the "coupled" mass corresponding to mostly "positional energy" , those "coupled" and "uncoupled" mass serve distinguish different nature of mediums where "emulsion" comes to play. Therefore, the gathering of bigger and bigger mass is driven by two forces! – the gravity and emulsion. The emulsion is well explained phenomenon in basic physics; it is not a subject of fantasy – dark energy though!

The pure positional energy, on the other hand, is not common on earth. However, it is not mystery and it is not dangerous if it is on a small scale. The author applaud European Particle accelerator, and assure the rest of us that black hole created in a human lab will not swallow earth! It is the game of quantity, a drop of water is not going to drown a giant elephant! The black hole is nothing but a compacted Gamma Ray nucleosynthesis. The author will address it over and over in this paper, because Gamma Ray Particles are the only suitable candidate for possessing greatest power in a very small package. As human, we are well aware of high power Gamma Ray during Gamma Ray Burst, however, the author can reasonable assure you if we pair liberated Gamma Ray/Particle together, its mighty kinetic energy will transform into mighty positional energy(γ ray nucleosynthesis or photosynthesis of *type*), a perfect candidate for dark matter/black hole.

The pairing process of solar light photons lays in the green photosynthesis- not a simple mechanic process. Similarly, the pairing process of Gama ray will not be a simple mechanic process either. The solar light photosynthesis is not a simple collision of two photons; it falls into a subject of organic chemistry discipline. The author is not sure about mechanic process of Gama ray photosynthesis; it might open up a brand new study subject of cosmos chemistry. The collision of two photons is probably not a way for coupling photons of any kind.

The universe energy is constant and it exists in two forms. Since we human have tools readily detect kinetic energy in distance, we know gamma ray is the matter possessing the most powerful kinetic energy in a very small package. We also know that a great bang is the most powerful explosion in universe. So gamma ray matches big blast!

Now, since the positional energy is not readily to be detected. However, we know that the energy is constant. The greatest positional energy must come from the greatest kinetic energy! What will be the matter possessing most positional energy? The only rationale answer for this is Gamma Ray nucleosynthesis or photosynthesis of γ type!

According to our recent cosmos findings, we know that there are black hole and dark matters which hold our visible mass together. The black hole/dark matters are regions / masses which possess most powerful positional energy in the universe. The author here safely says that black hole/dark matter is made from "non-liberated" gamma ray nucleosynthesis for most part. Since gamma ray particles are so small, it is very understandable that black hole doesn't take much space! The author's description is supported by recent discovery of Milky Way Black hole whose radius is no more than 6.25 light-hours! 6.25 light hours are plenty room for its dark matter holding something together that is sized 100,000–120,000 light-years in diameter which contains 100–400 billion stars, aka milky way, a home for human sapiens. The existence of gravitation center for a galaxy is a perfect example of basic physics, because everything will scatter away without a gravitation pull. Furthermore, the accelerated merging of galaxies/dark matter is directly resulted from force of gravity along with emulsion as author described in details earlier. They are all within the scope of basic physics. There may be "dark" energy as some people love to refer, the author, however, do not believe there are laws of physics "only" work in distant space where earth becomes a place of exception (excluding Disney world, of course). Unfortunately, the genius of icon often overlooks Newtown's head and his apple above, might not be a true genius in physics when mass worship overruns scientific scrutiny.

Gamma Ray, Big Bang and Accelerated Universe Expanding

It is the big bang that kicks off Gamma Ray first. The diffuse cosmic back ground

Gamma Ray is not originated during the period of Big Bang's nucleosynthesis (also known as primordial nucleosynthesis). The nucleosynthesis does happen, the process requires kinetic energy to be transformed into positional energy after particles start to mingle and gather around the dense regions. The mingle means when particles start to pair/orbit from each other that result in the orbiting power/energy. According to conservation of energy, the new orbiting power must come from somewhere other than from nothing. The only source is its kinetic energy. The evolution of a big bang simply means that some of its greatest kinetic energy starts to be captured into its form of orbiting energy/ gravity power during its cooling off stage, when galaxies start to be created. The destiny of big bang is another big bang. The gravity power along with power of emulsion (not energy of Disney world or space-time) will eventually collect enough mass and creates another new pre-big bang black hole!

The more mass is collected, the bigger a pre-big bang black hole grows, the stronger gravity becomes-directly evidenced by our observation of accelerated universe expanding! The accelerated speed is fuelled by growing mass/increasing gravity power! Our observed universe has passed a new born stage of post big bang, passed a maturing stage of numerous galaxies forming. Our observed universe is entering a big bang preparing stage. The accelerated expanding is a direct evidence of stronger gravity pull - larger and larger mass/ black hole are forming! For those who use accelerating universe as a reason doubting the big bang theory, the author can assure those doubters that the opposite is true!

The pre-big bang black hole is so enormous that it will dwarf our Milky Way's black hole as a drop of water vs. ocean. The enormous orbiting power that a black hole/dark matter possesses will eventually be released and transformed into kinetic energy, which presents itself in a spectacular gamma bomb show/ big bang explosion, which will dwarf gamma ray burst (GRB) as a 3 watts light bulb vs. the center of sun. However, the gamma ray burst (GRB) is not only a tiny rehearsal of a big bang, but also it proves to us that the real mechanism of spitting out gamma ray from a dark matter/black hole exists anytime!

We can reasonably believe that the pre-Big Bang Mass must contain mostly positional energy rather than kinetic energy, because soon it will kick off as an explosion. Accordingly, the explosion must be in a form of mostly kinetic energy, rather than in a form of positional energy. Before and during explosion, there is no logical reason to think that the universe space of any moment must be complete vacant. Here are the reasons. 1. The collection/ preparation stage of pre-Big Bang Mass only requires a quantifiable amount of mass/energy to trigger a big blast. There is no reason and need to stripe the space to a total void. [2] 2. The big bang is not the only event happening in the universe at any moment. There is mass that is constant exchanging from region to region regardless of status of a particular region. There are numerous big bangs. The various regions of universe are at different stage of mass/energy evolution process. Therefore, it further reduces possibilities of striping a space to a complete void even for a region that is ripened for a big blast.

My description of universe big blast is indirectly supported by general belief that primordial stage of clustering and merging exists. If the space were total void, the blast will send homogeneous Gamma Ray/Particles in a perfect sphere shape; there would be no area of density difference which is the key inducer for clustering and merging. The author reasons that non-void space which provides the density difference. Those density differences continue to exist before and during a big blast. Those density differences are necessity to induce clustering and merging which lead to dark matters, black holes and galaxies. Furthermore, those density differences spread out in space follow the mathematic rule of random. If the space were total void, it would have gradual changing appearance of cosmic ray spread in a sphere shape. Without those random existing of cosmos mass before and during big blast, we would not be here, neither is our Milky Way.

Gamma Ray and US

Gamma Ray is a good stuff. Gamma rays, electromagnetic radiation of an extremely high frequency / high energy photons, are produced by a number of astronomical processes in which very high-energy electrons are produced, that in turn cause secondary gamma rays via bremsstrahlung, inverse Compton scattering and synchrotron radiation. Gamma rays typically have frequencies above 10 exahertz (or >1019 Hz), and therefore have energies above 100 keV and wavelengths less than 10 picometers, which is less than the diameter of an atom. They can also be produced by the decay of atomic nuclei known as gamma decay. Gamma Ray also can be produced in our nuclear reaction etc.

There are 4 stages of utilizing energy in our human history. The First Stage is horse power stage. The humans mostly use animal power to supplement man's power. The second stage is Molecular Power stage. The humans mostly utilize fire power by burning fuels. The third stage is Nuclear Power stage. The humans mostly harvest energy from sub-atom nucleus. The fourth stage will be sub-nucleus Particle Power stage. The author applaud European Organization for Nuclear Research (CERN)'s effort to crackdown the sub-nucleus particles. The Large Hadron Collider (LHC), the world's largest and most powerful particle collider is the first step for humans entering the Era of Particle Power. The danger comes when particle power fall into a wrong hand. To protect our way of living and beloved human civilization, the international task to prevent particle power proliferation soon must be in its place! Comparing to pure kinetic energy, the pure positional energy is always hard 4to be detected in cosmos. The kinetic energy comes to us, while positional energy sits there. When this happens, the cosmos Gamma Ray serves perfect tool for us to notice the existence of positional energy/gravitation. The author's point is supported by phenomenon as gravitation lens. Furthermore, we shall be able to quantify the positional energy/gravitation by measuring gravitation lens effects carefully. The calculation will provide us not only the gravitation power but also the actual mass that invisible object possesses.

The earthlings are only one step from Gamma Ray communication. The Gamma Ray is much more powerful source that we can use to communicate cross vast cosmos distance. Humans are restricted by our physical forms which stop us from traveling in space. However, the electronic form of human in Gamma Ray will easily overcome the barriers that physical form of human face today.

Gamma Ray weapon will be the ultimate weapon dwarfing our current nuclear weapon systems as a catfish vs. Tyrannosaurus rex. Imagine that the GPS locate a target; Gamma Ray will evaporate it like mini big bang that hit a home run! The characteristic of gamma ray weapon is to leave no messy blood with a potential of recycling enemy's physical body and their weaponry. The nuclear weapon is more readily a source for γ ray nucleosynthesis. The era of γ particle warfare will become energy recycle enterprise. Not only the superior civilization send out cyber generals and solders (electronic form of human fighters) but also evaporate enemy and their arsenals into dark matter, take them home and sell them to a local utility company. The recycle of enemy energy is not a new concept. The very first war of human civilization involved capturing enemy soldiers and sold them to slave market to recycle their man power. In the era of particle power, enemy's sub-atom particle energy will be captured and recycled. The fight will eventually become bloodless and painless.

If there is any intelligent life exists outside earth, the Gamma Ray in space will be the main target to study. Gamma Ray is the most powerful energy carrier which is capable to travel the vastest distance. When the cosmos Gamma Rays reach us, some of them may carry information, or some Gamma Rays are alien life in their electronic form. Our human race is only one step below the ultimate level of terminal end of artificial intelligence. If there is superior alien life that exists somewhere in the universe, the only hope that they have to travel among galaxies or escape a big bang is to store their information in the gamma ray. The gamma ray is the most powerful ray available in the universe. Comparing to other types of ray, the Gamma ray might last longer and reach further. However, the artificial Gamma Ray with information may never escape a big bang. It is, though, worth our effort to collect cosmos Gamma Ray and look for clues that may mean something other than existing as a form of pure energy unit.

In a speed of light, all mass will break down into particles. The dream of human carrier reaching speed of light is just fantasy [2]. The human becomes unique and distinguish because of his/her personal memory and emotion that is associated with such memory. Those distinguished memories are nothing but an electronic form that is developed in our brain cells at first place. Once human medicine is able to upload and pack such electronic memory/emotion or data base into a light form, we will be able to travel in a speed of light. Today, you can beam your love letter to icloud. Tomorrow, who is to say that you could not beam yourself to icloud?

The only intelligent alien life that is superior to our human are the ones who master the Gamma Ray. The alien gamma rays wander in space, with hope to reach the receptors. They would not be able to complete their travel unless they find particular receiving receptors which allow them to down load. The colonization of cosmos will start in a form of Gamma Ray first, not a "May Flower" wooden boat. This sounds like a fiction, but it is science at its ultimate level. The particle physics is the main avenue to carry us to that level.

Other than for the potential of information storage, The Gamma Ray engine will be an ultimate power horse for inter stars/inter galaxies travel due to γ ray/fuels that are readily available in space.

The Gamma Ray will be an excellent detector for invisible mass. It is high energy, which travels fast and far, it will detect any positional energy nearby on its route due to the gravity lensing effects, or the bend of light. The Gamma Ray radar detector will be the ultimate detector available in the universe. There will be no stealth plane, or dark matters that can escape from its detection.

What We Are Made From

It has been said that we are made from star dust. This is not a quite correct answer. It is the inorganic matters that are made from star dust.

In about three minutes, The Big Bang nucleosynthesis produced most of the helium, lithium, and deuterium in the Universe, and perhaps some of the beryllium and boron. [10-12] The heavier elements are produced in stars through the process of nuclear fusion [13] see stellar nucleosynthesis for details. Isotopes such as lithium-6, as well as some beryllium and boron are produced in space through cosmic ray spallation. [14] This occurs when a high-energy proton strikes an atomic nucleus, causing large numbers of nucleons to be ejected. Elements heavier than iron are produced in supernovae through the r-process and in AGB stars through the s-process, both of which involve the capture of neutrons by atomic nuclei. [15] Elements such as lead formed largely through the radioactive decay of heavier elements. [16] Those elements exemplify what is called atom evolution.

The Organic Matters branch off from inorganic matters. The process requires light energy. The light energy could come from solar or earth itself. The most popular photosynthesis requires green chlorophyll pigments. In plants, these proteins are held inside organelles called chloroplasts, which are most abundant in leaf cells, while in bacteria they are embedded in the plasma membrane. The first photosynthetic organisms probably evolved early in the evolutionary history of life and most likely used reducing agents such as hydrogen or hydrogen sulfide as sources of electrons, rather than water. [17] Cyanobacteria appeared later, and the excess oxygen they produced contributed to the oxygen catastrophe, [18] which rendered the evolution of complex life possible.

The Organic Matters are mass possessing both kinetic energy and positional energy. To become alive and grow, the living beings require additional matter that possesses almost pure form of kinetic energy -light energy. However, the light energy is no alien to us. Neither is dark matter! Dark matter is just opposite from light energy, which possess positional energy mostly with very minimum kinetic energy. It is evidenced by universe background temperature. If there is temperature, there is kinetic energy.

Light has mass; it is evidenced by gaining slightly weight after photosynthesis. The current model of photosynthesis is perfect example that nature set up for us. When we enter the era of γ particle power, we will be able to utilize γ ray for photosynthesis.

Time and Space

Time is just a record of history that describes an evolution in process or movement in matter; if you were to stop time, you would see a snapshot or momentary freeze picture of matter physics -- a halted progress in matter/energy. The speed of matter/mass or their evolution can never stop; therefore, time will never stop. There is no disappearance of matter/mass/energy; there is only transformation of the engaging or dissembling with external or internal mass/energy. The Universe Mass and Energy are in an endless cycle, accordingly, time has no beginning and no ending. Space is just a void where matters float. The light can be bent, because the light is mass! But the Space is not a mass, neither is time! The space and time can never be bent!

Mathematic Odds and Déjà Vu

Since the universe matters are all made of same particles, whether they are gods particles or ghost particles, there is no mystery or magic force in our universe. It is all about odds in the universe where math comes to play. From the most primitive particles to the most sophisticated chemical structures, matter will never vanish. It just exists in different forms by pure odds. Given the infinite size of the universe, the probabilities are most likely infinite as well: "anything is possible." Therefore, at any moment, if you take a snapshot, you can always find materials/particles of same physical character at different places in other parts of universe simultaneously, which gives people a sensation of déjà vu. Similarly, when you look into a large crowd, I am sure some déjà vu is going to play right in front of your eyes -- two faces will appear to be identical. Two Identical person does not mean they are same person. Similarly, if you look deep in the universe, you might find another "you", the childhood memory, however, could be different.

Unify All Science Branches and Philosophy

The study of universe will eventually lead to great reunion of all science branches and philosophy as organic intelligence heads to its terminal stage.

RESULTS ANALYSIS

The universe is a complete and enclosed system, because there is nothing existing outside universe which may interferes universe matters/energy. The conservation of mass and energy perfectly apply universe as a unit. We can safely say that universe energy and mass are constant.

$$\mathbf{E}_{\text{bigbang}} = \mathbf{E}_k + \mathbf{E}_p$$

The $\mathbf{E}_k$ is the sum of kinetic energy presents in the current universe. $\mathbf{E}_p$ is the sum of positional energy presents in the current universe.

The big bang is a greatest event happened in the universe; it is an almost pure kinetic energy show. Soon after a big blast, some of its kinetic energy starts to transfer itself to the positional energy (γ ray nucleosynthesis). Some γ ray nucleosynthesis trapped at subatomic stage and evolves into dark matter (significant positional energy, little kinetic energy). Some γ ray nucleosynthesis goes on to form galaxies and us. Some γ ray remains free standing status and becomes cosmos background ray.

The cosmos background ray may have diversified sources, due to supernova. It may also have organic intelligence originating, though the artificial gamma ray would be very little to none at its scale vs. a great bang or a supernova.

DISCUSSION

Our universe is an amazing place, but our milky way is full of amusing theorists. The author in this paper tries to return things to its practical basics and reveal the mechanic truth. The common sense/basic physics laws works out for our earth, and it will work for our universe as well. Surprisingly to some genius, our earth is part of the universe, not a place of exclusion!

In this paper, the author uses basic common sense to pave a road to disclose our universe mysteries by mapping its energy evolution. The conservation of energy/mass perfectly apply universe as a unit, since there is nothing existing outside universe which may interferes universe matters / energy. Therefore, the author postulates 1. The common sense/basic physics perfectly predicts and reasons the physical necessity of existence of dark matter/black hole for every galaxy; 2. The conservation of energy dictates that a big blast/bang kinetic energy must come from a significant source of positional energy. In the universe, nothing is more qualified for a significant source of positional energy than a black hole/dark matter gathering. Those views of kinetic energy that can come from nothing/void are not deserved for a scientific debate. 3. The conservation of energy perfectly explains the reasons that dark matter/black hole must be invisible. 4. The significance of the universal cosmic gamma ray links gamma ray to a big blast/bang.

Furthermore, the recent discovery of gamma ray burst (GRB) from a black hole/dark matter directly supports the hypotheses of a pre-big bang mass which is dark matter/black hole in its nature. The GRB is not only a rehearsal of a big blast, but also it provides a sneak preview that dark matter has the mechanism to spit out gamma ray before its show time-big bang!

Given the significance of Gamma Ray, the author labels it as a terminal stage of artificial intelligence in utilizing energy. The first 3 stages of utilizing energy in our human history are 1. Horse power stage 2. Molecular Power stage e.g. burning fuels. 3. Nuclear Power stage. This research paper promotes the study of particles, especially gamma ray, which demands imminent attention from advanced civilizations.

Under dictations of the laws of basic physics, the author further reasons what dark matter/black hole is made of- γ ray nucleosynthesis or photosynthesis γ type. Furthermore, the author describes the two forces that drive galaxies merge in space – gravity and emulsion. Finally, the author projects the mode of mass evolution in our universe by analyzing its energy composition. In consideration of infinite size of the universe, the author postulates numerous big bangs exist in our universe with endless mass/energy cycles.

REFERENCES

1. Valencia, G. (1992). "Anomalous Gauge-Boson Couplings At Hadron Supercolliders". AIP Conference Proceedings 272: 1572–1577.

2. Zong, E. (2014) One universe, endless cycles.

3. Zong, E. (2011) Understanding of Universe, Time, Odds and Environment.

4. Kumar, K. N. P.; Kiranagi, B. S.; Bagewadi, C. S. "Hawking Radiation-A Augmentation Attrition Model".

5. Ghez, A. M.; Salim, S.; Hornstein, S. D.; Tanner, A.; Lu, J. R.; Morris, M.; Becklin, E. E.; Duchêne, G. (May 2005). "Stellar Orbits around the Galactic Center Black Hole". The Astrophysical Journal 620 (2): 744–757. ar Xiv:astro-ph/0306130. Bibcode: 2005ApJ...620..744G. doi:10.1086/427175.

6. Aharonian, F.; Akhperjanian, A.; Barrio, J.; Bernlohr, K.; Borst, H.; Bojahr, H.; Bolz, O.; Contreras, J.; Cortina, J.; Denninghoff, S.; Fonseca, V.; Gonzalez, J.; Gotting, N.; Heinzelmann, G.; Hermann, G.; Heusler, A.; Hofmann, W.; Horns, D.; Iserlohe, C.; Ibarra, A.; Jung, I.; Kankanyan, R.; Kestel, M.; Kettler, J.; Kohnle, A.; Konopelko, A.; Kornmeyer, H.; Kranich, D.; Krawczynski, H.; Lampeitl, H. (2001). "The TeV Energy Spectrum of Markarian 501 Measured with the Stereoscopic Telescope System of HEGRA during 1998 and 1999". The Astrophysical Journal 546 (2): 898. Bibcode:2001ApJ...546..898A. doi:10.1086/ 318321.

7. NASA'S Fermi Telescope Unveils a Dozen New Pulsars http://www. nasa.gov/mission_pages/GLAST/news/dozen_pulsars.html.

8. Cosmos Online – New Kind of pulsar discovered (http://www. cosmosmagazine.com/news/2260/new-kind-pulsar-discovered).

9. BBC Gamma Ray Burst.

10. Croswell, Ken (1991). "Boron, bumps and the Big Bang: Was matter spread evenly when the Universe began? Perhaps not; the clues lie in the creation of the lighter elements such as boron and beryllium". New Scientist (1794): 42. Archived from the original on 7 February 2008. Retrieved 14 January 2008.

11. Copi, Craig J.; Schramm, DN; Turner, MS (1995). "Big-Bang Nucleosynthesis and the Baryon Density of the Universe". Science 267 (5195): 192–99. arXiv:astro-ph/9407006. Bibcode: 1995Sci...267..192C. doi:10.1126/science.7809624. PMID 7809624.

12. Hinshaw, Gary (15 December 2005). "Tests of the Big Bang: The Light Elements". NASA/WMAP. Archived from the original on 17 January 2008. Retrieved 13 January 2008.

13. Hoyle, F. (1946). "The synthesis of the elements from hydrogen". Monthly Notices of the Royal Astronomical Society 106: 343–83. Bibcode:1946MNRAS.106..343H. doi:10.1093/mnras/106.5.343.

14. Knauth, D. C.; Knauth, D. C.; Lambert, David L.; Crane, P. (2000). "Newly synthesized lithium in the interstellar medium". Nature 405 (6787): 656–58. doi:10.1038/35015028. PMID 10864316.

15. Mashnik, Stepan G. (2000). "On Solar System and Cosmic Rays Nucleosynthesis and Spallation Processes". arXiv: astro-ph/0008382 astro-ph.

16. Kansas Geological Survey (4 May 2005). "Age of the Earth". University of Kansas. Retrieved 14 January 2008.

17. Olson JM (May 2006). "Photosynthesis in the Archean era". Photosyn. Res. 88 (2): 109–17. doi:10.1007/s11120-006- 9040-5. PMID 16453059.

18. Buick R (August 2008). "When did oxygenic photosynthesis evolve?". Philos. Trans. R. Soc. Lond., B, Biol. Sci. 363 (1504): 2731–43. doi:10.1098/rstb.2008.0041. PMC 2606769. PMID 18468984.

Chapter 9

ANALYSIS OF REACTIVITY - INITIATED ACCIDENT FOR CONTROL RODS EJECTION

H. Mansour[1], Hend M. Saad[2], M. Aziz[2]

[1]Physics Department, Faculty of Science, Cairo University, Giza, Egypt
[2]Nuclear and Radiological Regulatory Authority, Ahmed Al Zomor St., Nasr City, Cairo, Egypt

ABSTRACT

Understanding of the time-dependent behavior of the neutron population in a nuclear reactor in response to either a planned or unplanned change in the reactor conditions is of great importance to the safe and reliable operation of the reactor. In the present work, the point kinetics equations are solved numerically using the stiffness confinement method (SCM). The solution is applied to the kinetic equation in the presence of different types of reactivities, and is compared with other method. This method is, also used to analyze reactivity induced accidents in two reactors. The first reactor is fueled by uranium and the second is fueled by plutonium. This analysis presents the effect of negative temperature feedback with the addition positive reactivity of the control rods to overcome the occurrence of the control rod ejection accident and damaging of the reactor. Both the power and the temperature pulse following the reactivity- initiated accidents are calculated. The results are compared with previous works and satisfactory agreement is found.

INTRODUCTION

The reactivity- initiated accident is a nuclear reactor accident that involves inadvertent removal of a control element from an operating reactor, thereby causing a rapid power excursion in the nearby fuel elements and temperature. The postulated scenarios for reactivity initiated accidents are therefore focused on a few events, which result in exceptionally large reactivity excursions, and therefore are critical to fuel integrity. In reference[2] model, the reactivity-initiated accident is considered to be due to negative temperature feedback. In the present work, we consider the reactivity accident to be due to negative temperature feedback with the addition positive reactivity of the control rods to prevent such accidents of the control rods ejection and damaging of the reactor. We analyzed accidents in different types of reactors, e.g. modular high temperature gas cooled reactor design like HTR-M, and modular fast reactor design like PRISM, using the stiffness confinement method for solving the kinetics equations. The stiffness confinement method SCM is used to solve the kinetics equations, and overcome the stiffness problem in reactor kinetics[1]. The idea is based on the observation that the stiffness characteristic is present only in the time response of the prompt neutron density, but not in that of the delayed neutron precursors. The method is therefore devised to have the stiffness decoupled from the differential equation for precursors and is confined to the one for prompt neutrons, which can be solved[1]. Numerical examples of applying the method to a variety of problems are given. The method is also used to analyze the reactivity induced accidents in two reactors data, the modular high temperature gas cooled reactor (HTR-M) which is fueled by uranium, and modular fast reactor design (PRISM) which is fueled by plutonium. In the next sections, we discuss the mathematical method; present the results and discussion, and give the conclusion.

MATHEMATICAL METHOD

The stiffness confinement method is used to overcome the stiffness problem in reactor kinetics for solving the point kinetics equations. The idea is based on the observation, that the stiffness characteristic is present only in the response time of the prompt neutron density, but not in that of the delayed neutron precursors. The method is, therefore, devised to have the stiffness decoupled from the differential equations for the precursors and confine it to the one for the prompt neutrons, which can be analytically solved[1]. The point kinetics equations are a system of coupled ordinary differential equations, whose solution give the neutron density and delayed neutrons precursor concentrations in a tightly coupled reactor as a function of time. Typically, these equations are solved using the reactor model with at least six delayed precursor groups, resulting in a system consisting of seven coupled differential equations. Obtaining accurate results is often problematic, because the equations are stiff with many techniques, where very small time steps are used. These equations take the following form with an arbitrary reactivity function[3, 4]:

$$\frac{dn(t)}{dt} = \frac{\rho(t)-\beta}{\Lambda} n(t) + \sum_{i=1}^{6} \lambda_i C_i(t)$$

(1)

$$\frac{dC_i(t)}{dt} = \frac{\beta_i}{\Lambda} n(t) - \lambda_i C_i(t)$$

(2)

where: n(t) is the time-dependent neutron density, or (power or neutron flux) all units are (MW) as power unit; $C_i(t)$ is the ith group delayed neutrons precursor concentration or delayed neutrons emitter population or precursor density ("latent-neutron" density or latent power; same units as in the power); i is the number of precursor group; $\rho(t)$ is the time-dependent reactivity; β_i is i^{th} group delayed neutrons fraction, and $\beta = \Sigma_i \cdot \beta_i$, is the total delayed neutrons fraction. In addition, Λ is the neutron generation time (s), and λ_i is decay constant of the ith group delayed neutrons emitters (s^{-1}).

Introducing a set of "Reduced" precursor density functions $\hat{C}i$ (t) and neutron density n (t), through the following equation[1]:

$$C_i(t) = \hat{C}_i(t) \exp[\int_0^t u(t')\, dt']$$

(3)

Defining two auxiliary functions w (t) and u (t), as in Eqs. (4) and (5):

$$w(t) = \frac{d}{dt} \ln n(t)$$

(4)

The function w (t) is defined in Eq. (9) below and provides the key mechanism of the SCM. The function u (t), however, has nothing to do with stiffness decoupling and is not really required theoretically. Since an exponential behavior is often characteristic for the first, order differential equations, however, a proper choice of u (t) may make $\hat{C}_i$ (t) vary more slowly in time and thus expedite the numerical calculation. Choose the following u (t) [1]:

$$u(t) = \frac{d}{dt} \ln S(t)$$

(5)

Where, S (t) is defined by Eq. (7) as the sum over all $\lambda_i \cdot C_i$ (t). We can rewrite Eqs. (1) and (2) as follows[1]:

$$\frac{d\hat{C}_i(t)}{dt} = \left[\frac{\beta_i}{\Lambda\, w(t) + \beta - \rho(t)} \right] \sum_{i=1}^{6} \lambda_i \hat{C}_i(t)$$

$$-\left[u(t) + \lambda_i \right] \hat{C}_i(t)$$

(6)

$$S(t) = \left[\sum_{i=1}^{6} \lambda_i \hat{C}_i(t) \right] \exp\left[\int_0^t u(t') dt' \right]$$

(7)

And,

$$\frac{dn(t)}{dt} = \frac{\rho(t) - \beta}{\Lambda} n(t) + S(t)$$

Suppose that, it is always possible to express:

$$n(t) = \exp[\int_0^t w(t') dt']$$

(8)

and rewrite Eq. (1) as:

$$n(t) = \frac{\sum_{i=1}^{6} \lambda_i C_i(t)}{\left(w(t) + \left[\dfrac{\beta - \rho(t)}{\Lambda} \right] \right)}$$

(9)

Eqs. (6)- (9), form the complete set of kinetics equations for the SCM. The initial conditions are satisfied to be:

$$u(0) = 0$$

(10a)

$$w(0) = \frac{\rho(0)}{\Lambda}$$

(10b)

$$n(0) = n_0$$

(10c)

And;

$$\hat{C}_i(0) = \frac{n_0 \beta_i}{\Lambda \lambda_i}$$

(10d)

By using the initial conditions, we can obtain the numerical solution of the equations. We first start by setting w and u in Eq. (6) at their initial values and solve Eq. (6) for $\hat{C}_i$ by discretizing the equation in t. Having obtained $\hat{C}_i$ we calculate S (t) with Eq. (7). Then we use Eq. (4) to re-evaluate w(t), plug it back into Eq. (6) , and repeat the process until w converges (requiring 50 iterations). Calculation for the current time step is finished with an evaluation of the output value of w and u via Eqs.(4) and (5). Afterward, we predict the input values of w and u for the next time step by a linear extrapolation from their output values in the previous and current time steps, and repeat the whole process of calculation for the next time step. It should be emphasized that time step, there is iteration to convergence on w, but no iteration for the function u, because u is not required by the theory of the SCM, and is in principle, an arbitrary independent function chosen only to expedite the computation. A computer program is designed with programming language FORTRAN, and MATLAB code to solve the above equations numerically using Runge - Kutta method for the above differential equations and the output power and temperature are determined under different input reactivities.

It is assumed that, the reactor has a negative temperature coefficient of reactivity α ($\alpha > 0$), when a large step reactivity ρ_0 ($\rho_0 > \beta$) is inserted. Consider the temperature feedback, the real reactor reactivity is:[3, 4]

$$\rho(t) = \rho_0 - \alpha \left[T(t) - T_0 \right] \tag{11}$$

Then, the derivative of Eq. (11) with respect to time (t) is:

$$\frac{d\rho(t)}{dt} = -\alpha \frac{dT(t)}{dt} \tag{12}$$

Where, T (t) and T_0 are the reactor temperature, and initial temperature of the reactor, respectively. After the large reactivity ρ_0 is inserted into the reactor, the power responds quickly and the adiabatic mode can be used for the calculation of reactor temperature.[3, 4]

Then, the derivative of the temperature w.r.to time can be given as follows:

$$\frac{dT(t)}{dt} = K_c \, n(t) \tag{13}$$

Where, K_c is the reciprocal of thermal capacity of reactor. Substituting Eq. (12) into Eq. (13) results in the following:

$$\frac{d\rho(t)}{dt} = -\alpha \, K_c \, n(t) \tag{14}$$

NUMERICAL SOLUTION

The numerical solution of the point kinetics equations is based on SCM .The results are compared against other result, which obtained with other method. The other method is highly accurate, but there are vary widely in there complexity of implementation [1].

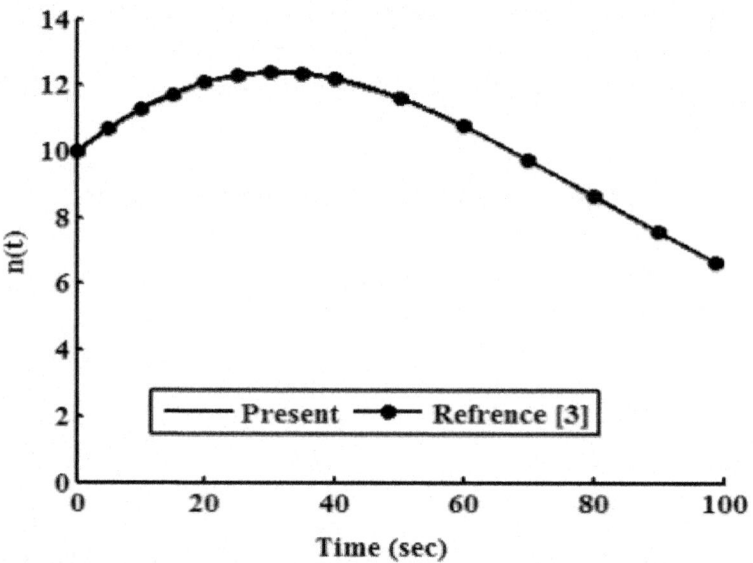

Figure 1: Neutron density as a function of time at $\rho_0=0.2\beta$

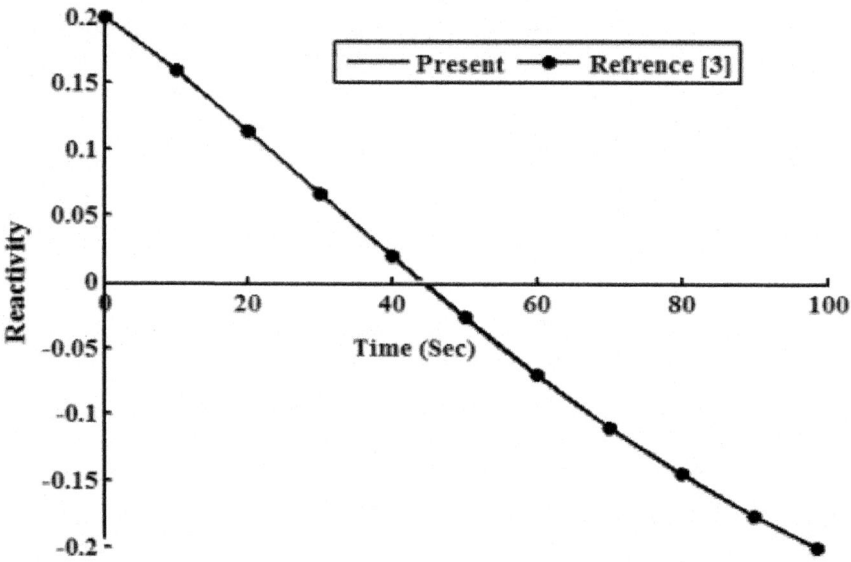

Figure 2: Reactivity as a function of time at $\rho_0=0.2\beta$

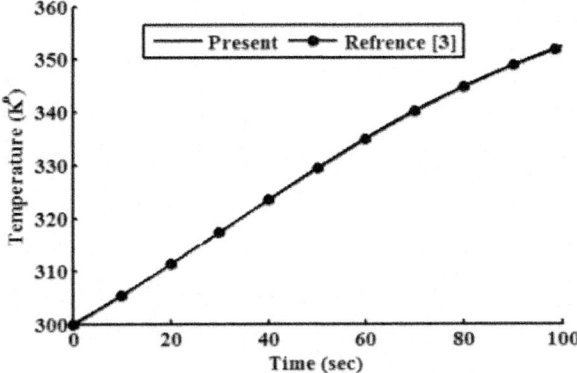

Figure 3: Temperature as a function of time at $\rho_0 = 0.2\beta$

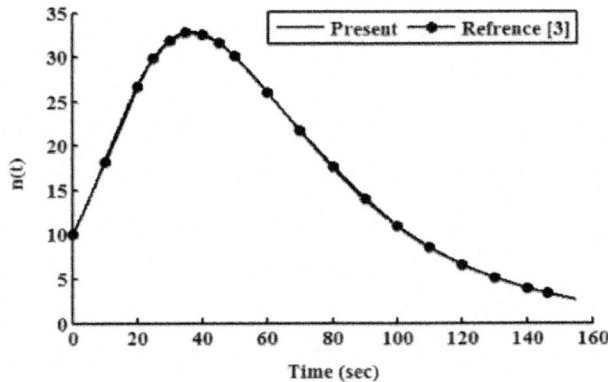

Figure 4: Neutron density as a function of time at $\rho_0 = 0.5\beta$

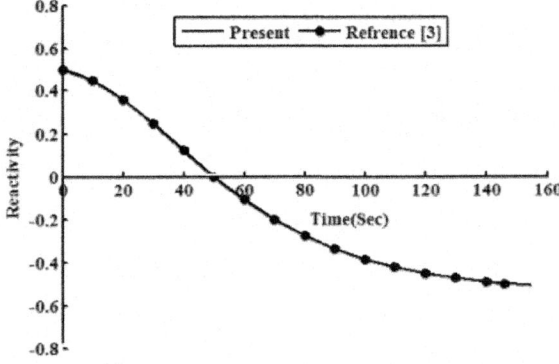

Figure 5: Reactivity as a function of time at $\rho_0 = 0.5\beta$

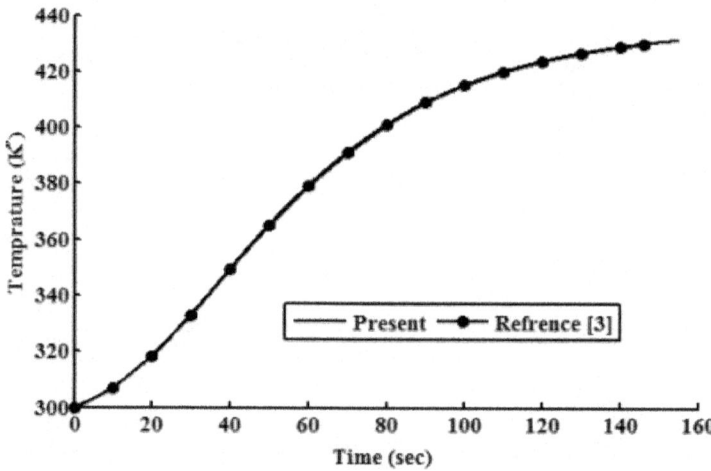

Figure 6: Temperature as a function of time at ρ_0=0.5β

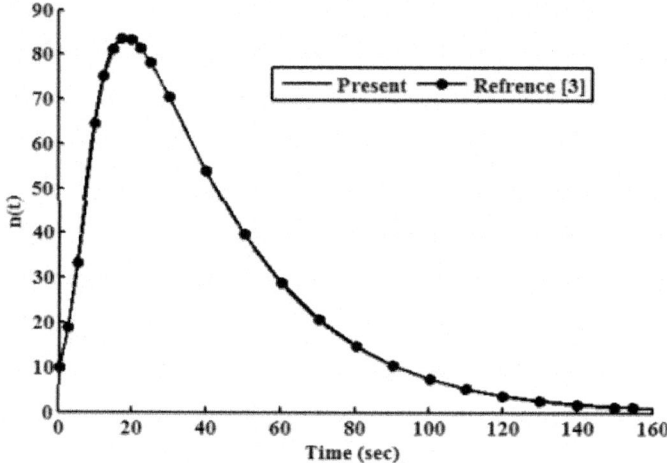

Figure 7: Neutron density as a function of time at ρ_0=0.8β

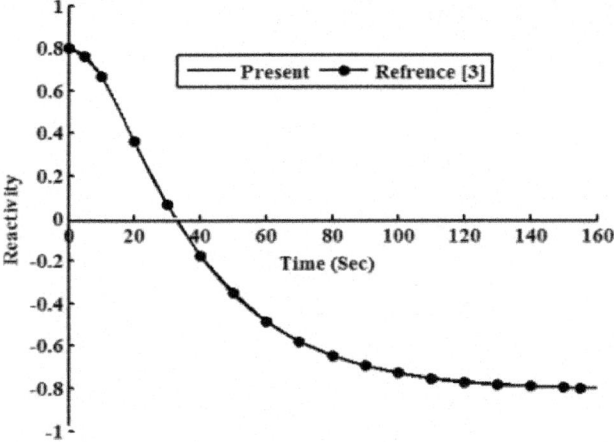

Figure 8: Reactivity as a function of time at $\rho_0 = 0.8\beta$

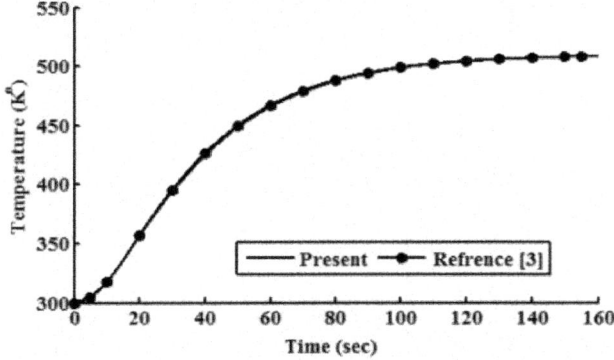

Figure 9: Temperature as a function of time at $\rho_0 = 0.8\beta$

Considering an example ^{235}U as a fissile material under large step reactivity, taking three initial reactivities inserted into the reactor. It is assumed that, the input parameters are: $\lambda_i(s^{-1}) = (0.0127, 0.0317, 0.155, 0.311, 1.4, 3.87)$, $\beta_i =$ (0.000266, 0.001491, 0.001316, 0.002849, 0.000896, 0.000182), $\Lambda=0.0001(s)$, $\beta_{tot} = 0.0065$, $K_c = 0.05$ K/ (MW s), and $\alpha = 5\times10^{-5}$ (K^{-1}). In this case, $\rho_0 = 0.2\beta$ and 0.5β and 0.8β for $t \geq 0$ are used into the reactor, which is operating in critical state with initial power 10 (MW). The relation between time, reactivity, neutron density, and temperature using the numerical solution by the SCM, which are compared with the work of[3], is presented in tables (1, 2, and 3). The results indicate that the present model solutions are in good agreement with other works as shown in following figures (1-8). The iteration in computing is used for repeating the process until w and u converge (requiring approximately 100 iterations) to get step reactivity insertion with accurate results which are compared with other calculations [1, 3].

Table 1: Neutron Density, Precursor Density, Reactivity and Temperature are functions of time at initial reactivity 0.2β solved

Time (sec)	Reactivity ($)		Neutron Density (MW)		Precursor Density		Temperature (K°)	
	Present Work	Reference[3]	Present Work	Reference[3]	Present Work	Reference[3]	Present Work	Reference[3]
0.0	0.2	0.2	10.0	10.0	6718	6717.479	300	300
10.0	0.1589	0.158976	11.28	11.270597	7962	7960.6118	305.3	305.333
20.0	0.1139	0.113885	12.11	12.094137	8999	8999.4785	311.2	311.195
30.0	0.06659	0.066610	12.4	12.402006	9720	9720.1787	317.3	317.341
40.0	0.0191	0.019115	12.22	12.220168	10060	10064.441	323.5	323.515
50.0	-0.02687	-0.026878	11.63	11.639050	10030	10034.852	329.5	329.494
60.0	-0.07006	-0.070057	10.78	10.777035	9682	9682.0400	335.1	335.107
70.0	-0.1096	-0.109569	9.751	9.750800	9083	9083.3223	340.2	340.244
80.0	-0.145	-0.144980	8.659	8.658279	8323	8322.8449	344.8	344.847
90.0	-0.01762	-0.176181	7.573	7.572253	7477	7477.1450	348.9	349.04
98.7	-0.2	-0.2	6.673	6.671694	6721	6721.2397	352	352

Table 2: Neutron Density, Precursor Density, Reactivity and Temperature are functions of time at initial reactivity 0.5β

Time (sec)	Reactivity ($)		Neutron Density (MW)		Precursor Density		Temperature (K°)	
	Present Work	Reference [3]	Present Work	Reference [3]	Present Work	Reference [3]	Present Work	Reference [3]
0.0	0.5	0.5	10.0	10.0	4198	4198.4243	300	300
10.0	0.4464	0.446517	18.21	18.195158	8476	8467.9072	307.0	306.953
20.0	0.3594	0.359572	26.75	26.726191	14390	14381.679	318.3	318.258
30.0	0.2453	0.245386	31.89	31.95273	20220	20213.748	333.1	333.100
40.0	0.1200	0.120072	32.59	32.597210	24080	24083.318	349.4	349.391
50.0	0.00126	0.001244	30.13	30.103104	25300	25304.049	365.2	365.162
60.0	-0.1096	-0.109559	26.08	26.081541	24290	24294.025	379.2	379.243
70.0	-0.2015	-0.201433	21.68	21.698225	21880	21884.133	391.2	391.186
80.0	-0.2768	-0.276824	17.58	17.578548	18840	18841.465	401	400.987
90.0	-0.3373	-0.337342	13.99	13.988706	15710	15704.299	408.9	408.854
100.0	-0.3854	-0.385198	11	10.995258	12790	12785.422	415.1	415.076
110.0	-0.4225	-0.422648	8.578	8.567273	10240	10231.454	419.9	419.944
120.0	-0.4516	-0.451736	6.641	6.633668	8093	8084.229	423.7	423.726
130.0	-0.4743	-0.474206	5.118	5.113089	6335	6327.5962	426.6	426.647
140.0	-0.4916	-0.491496	3.927	3.927845	4923	4917.8364	428.9	428.895
146.1	-0.5	-0.5	3.345	3.339961	2409	4205.6318	430	430

Table 3: Neutron Density, Precursor Density, Reactivity and Temperature are functions of time at initial reactivity 0.8β

Time (sec)	Reactivity (S) Present Work	Reactivity (S) Reference [3]	Neutron Density (MW) Present Work	Neutron Density (MW) Reference [3]	Precursor Density Present Work	Precursor Density Reference [3]	Temperature (K°) Present Work	Temperature (K°) Reference [3]
0.0	0.8	0.8	10.0	10.0	1679	1679.3699	300	300
10.0	0.6661	0.666886	64.74	64.557487	18210	18124.689	317.4	317.305
20.0	0.3639	0.36352	83.3	83.320236	44500	44500.184	356.7	356.699
30.0	0.06437	0.064528	70.43	70.490250	55350	5534.738	395.6	395.611
40.0	-0.1744	-0.174378	53.9	53.917503	53140	53148.172	426.7	426.699
50.0	-0.3535	-0.353502	39.77	39.60925	45170	45172.699	450	449.95
60.0	-0.4844	-0.484405	28.81	28.824406	35910	35915.492	467	466.973
70.0	-0.5788	-0.78847	20.73	20.703527	27440	27438.424	479.2	479.250
80.0	-0.6465	-0.646493	14.78	14.789428	20440	20440.443	488	488.044
90.0	-0.6947	-0.694731	10.53	10.528523	14980	14977.864	494.3	494.315
100.0	-0.7290	-0.729082	7.484	7.478508	10850	10854.269	498.8	49.774
110.0	-0.7534	-0.753380	5.309	5.304107	7819	7806.8105	501.9	501.939
120.0	-0.7707	-0.770631	-0.770631	3.758143	5595	5585.821	504.2	504.182
130.0	-0.7828	-0.782857	2.665	2.660907	3989	3982.2925	505.8	505.771
140.0	-0.7915	-0.791503	1.886	1.83123	2832	283.9358	506.9	506.895
150.0	-0.7976	-0.797624	1.355	1.332234	2008	2010.3282	507.7	507.691
155.0	-0.8	-0.8	1.122	1.116575	1696	1687.1426	508	508.002

ANALYSIS OF REACTIVITY- INITIATED –ACCIDENT

Reactivity- Initiated Accident

Reactivity- initiated accident involves an unwanted increase in fission rate and reactor power. The power increase may damage the reactor core, and in very severe cases, even lead to the disruption of the reactor. The immediate consequence of reactivity- initiated accident is a fast rise in fuel power and temperature. The power excursion may lead to failure of the nuclear fuel rods and release of radioactive material into primary reactor coolant. In this study, a new computer program has been developed for simulating the reactor dynamic behavior during reactivity induced transients, and it has been used for the analysis of specified reactivity - initiated accidents in several cases. We introduce the two model reactors with system parameters that are characteristic for modular high temperature gas-cooled reactor design like HTR-M[6], and modular fast reactor design like PRISM[7]. For simplicity, we refer to the input dates of two reactors (HTR-M and PRISM) in tables (4, 5, and 6). For the delayed neutron parameters, it is assumed that, HTR-M is fuelled by ^{235}U and PRISM by^{239}Pu as fissile nuclides. The dynamic equations (15:21) for the two models are the conventional point reactor kinetic equations in combination with a linear temperature feedback for the reactivity, an adiabatic heating of the core after loss of cooling [2], where Eq. (17 a) may be modified to add positive control rods reactivity as:

$$\frac{dn(t)}{dt} = \frac{\rho_{net}(t) - \beta}{\Lambda} n(t) + \sum_{i=1}^{6} \lambda_i C_i(t) \qquad (15)$$

$$\frac{dC_i(t)}{dt} = \frac{\beta_i}{\Lambda} n(t) - \lambda_i C_i(t) \qquad (16)$$

$$\rho_{net}(t) = \rho_{feed}(t) + \rho_{ext}(t)$$
$$\rho_{feed} = feedback\ reactivity \qquad (17)$$

$$\rho_{feed} = -\alpha(T(t) - T_0) \qquad (18a)$$

$$\rho_{ext} = \rho_{CR} = \rho_{cr1}$$
$$or\ \rho_{cr2}\ or\ \rho_{cr3}\ or\ \rho_{cr4} \qquad (18b)$$

$$\rho_{ext} = external\ reactivity$$
$$= control\ rods\ reactivity$$

$$\rho_{net}(t) = -\alpha(T(t) - T_0 + \rho_{CR} \tag{19}$$

$$\frac{dT(t)}{dt} = \frac{1}{c} n(t) \tag{20}$$

Where, n (t) = reactor power (MW), ρ_{net} (t) = is the time-dependent reactivity function, ρ_{CR} = Addition positive reactivity of the control rods, β = total delayed neutron fraction, $\beta = \Sigma_i \beta_i$. β_i = Delayed neutrons faction of ith group. Λ = neutron generation time (sec), λi decay constant of ith group delayed neutrons emitters (sec)$^{-1}$, C_i (t) =delayed neutrons emitter population (in power units), α = temperature coefficient of reactivity (K^{-1}).

In the equation of total reactivity ρ_{net} (t), the additional positive reactivity of the control rods ρ_{CR} has four cases to prevent the control rods ejection accident as:

The input parameters of the kinetic equations of two reactors with different fissile materials are shown in tables 4: 6.

$$\rho cr1 = \rho 1 = 0, \rho cr2 = \rho 2 = (\beta/2),$$
$$\rho cr3 = \rho 3 = (0.8\beta), \rho cr4 = \rho 4 = (\beta) \tag{21}$$

Table 4: ^{235}U (Thermal Neutrons

λ_i(sec^{-1})	0.0124	0.0305	0.111	0.301	1.14	3.01
β_i	0.000215	0.001424	0.001274	0.002568	0.0007485	0.0002814
β_{tot}=0.0067			Λ=1.00E-4(sec)			

Table 5: ^{239}PU (Fast Neutrons)

λ_i(sec^{-1})	0.0129	0.0311	0.134	0.331	1.26	3.21
β_i	7.6E-005	5.6E-004	4.32E-004	6.56E-004	2.06E-004	7.00E-005
β_{tot}=0.0020			Λ=1.00E-7(sec)			

Table 6: Adiabatic Inherent Shutdown Data for Two Model Reactors

Types of Reactors	P$_0$(MW)	c(MJ/K)	α(K^{-1})
HTR-M	200.00	100.00	2.2E-005
PRISM	470.00	200.00	9.00E-006

Reactivity Evaluation

The reactivity of one, two and three control rods worth are calculated based on the assumptions of relating the control rod worth by the delayed neutron fraction β. We assumed that, the ejection of one, two and three rods could induce positive reactivity as indicated in Table 7 for each type of reactors, in the two models.

Table 7: Additional Positive Reactivity of Control Rods Insertion

No. of control rods	ρ (in S) for U^{235}	ρ (in S) for PU^{239}
1	0.5	0.5
2	0.8	0.8
3	1.00	1.00

RESULTS AND DISCUSSION

First Reactor (PRISM Reactor)

PRISM Reactor is assumed to be critical at the zero power condition, and the limited value of time (sec) on x axis is 300 (sec). Reactivity is also added step by step. The control rod insertion increases the thermalization of neutrons, results in a positive reactivity addition. Control rod insertion requires a certain driving force. The driving forces on the control rods in the reactors are the buoyancy from the fuel material and the supporting force from the control system of the reactor. If the control system should lose the support of control rods or control rods should break, control rods would be flown out of the reactor. Thus, in PRISM reactor, accidental insertions can result from the ejection of control rod drive, and/or control rod control system or operator error. Power transients are shown in Figures (10, 11) for one and two, and three control rod ejections, respectively. As can be seen, three rods are ejected; a large power pulse is generated about 83.8085 times of the initial value of power in a very short time. This is because; the accident is reactivity accident. However, the fuel temperature is stabilized at approximately 1435 (K) at the maximum as shown in Figure 12. Even, three control rods are inserted; the maximum temperature is not exceeding 1502 (K) as shown in Figure 12.

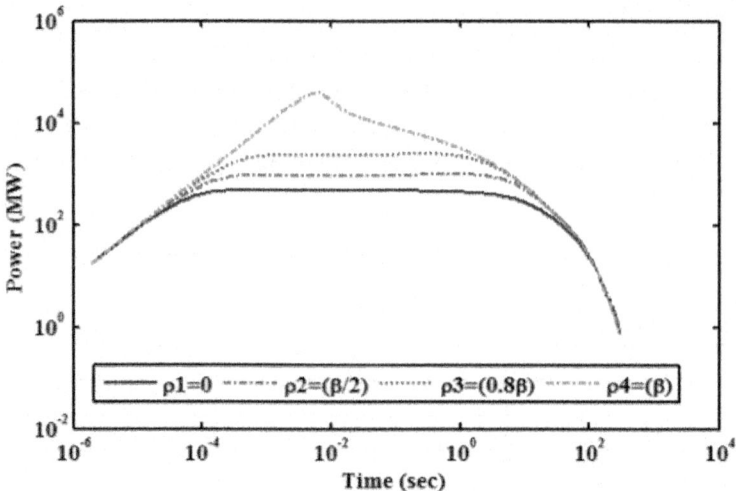

Figure 10: Shows Power (MW) Transient at Zero Power Condition with Different Values of Positive Reactivity of Control Rods Ejection for PRISM Reactor

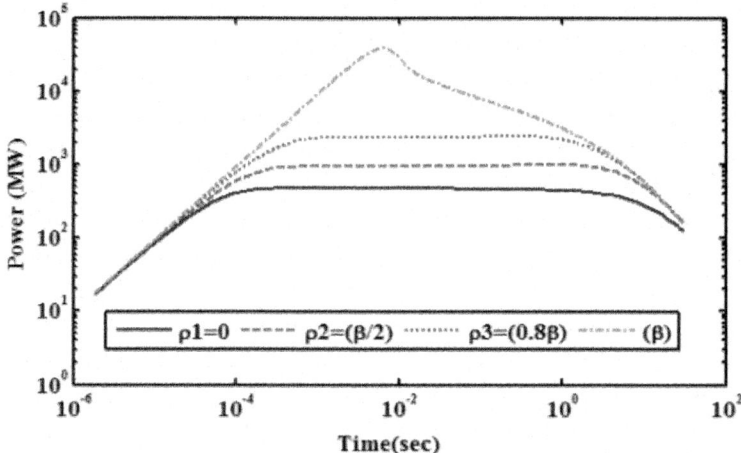

Figure 11: Shows Power (MW) Transient at Full up to 30s with Different Values of Positive Reactivity of Control Rods Ejection for PRISM Reactor

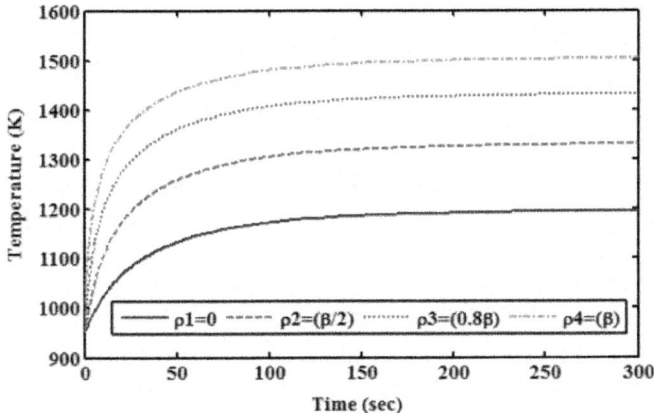

Figure 12: Shows Temperature (K) During the Transients at Zero Power for PRISM Reactor

Second Reactor (HTR-M Reactor)

HTR-M Reactor is assumed to be critical at the zero power condition, and the limited value of time (sec) on x axis is 300 (sec). Reactivity is also added step by step as explained above in PRISM reactor.

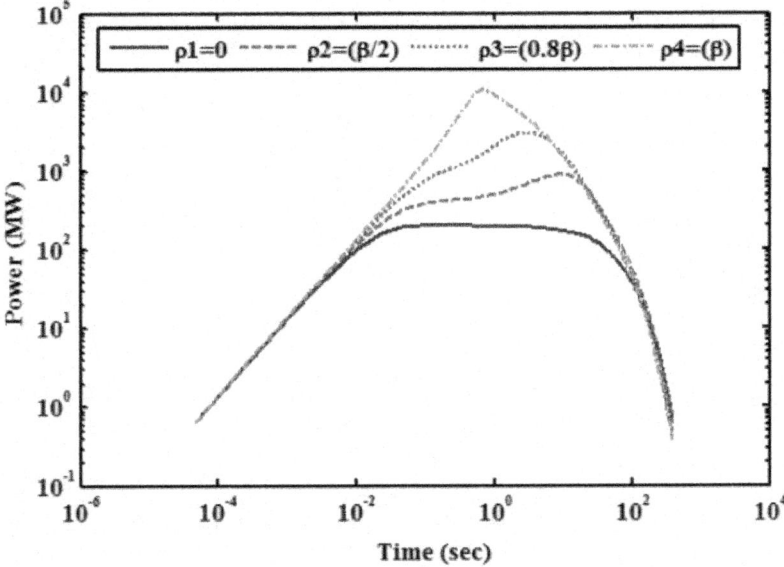

Figure 13: Shows Power (MW) Transient at Zero Power Condition with Different Values of Positive Reactivity of Control Rods Ejection for HTR-M Reactor

The Power transients for one, two, and three control rods ejection are shown in Figures (13, 14), respectively. As can be seen, three rods are ejected; a large power pulse is generated about 53.45 times from the initial value of power in very short time. This is because; the accident is reactivity accident. However, the fuel temperature is stabilized at approximately 916.8 (K) at the maximum as shown in Figure 15. Even, three control rods are inserted; the maximum temperature is not exceeding 983.1 (K) as shown in Figure 15.

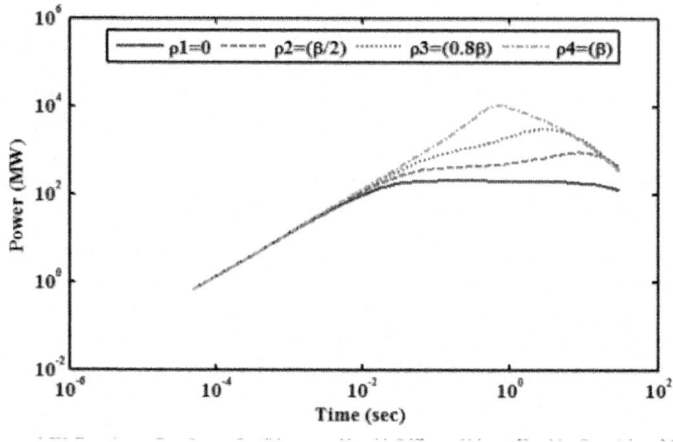

Figure 14: Shows Power (MW) Transient at Zero Power Condition up to 30s with Different Values of Positive Reactivity of Control Rods Ejection for HTR-M Reactor

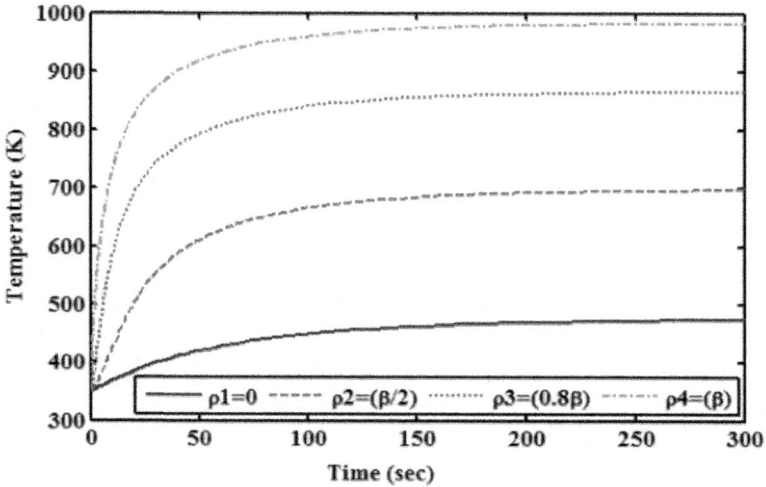

Figure 15: Shows Temperature (K) During the Transients at Zero Power for HTR-M Reactor

CONCLUSIONS

A computer program is designed to solve the point reactor dynamics equations using the stiffness confinement method (SCM) and different input reactivity. The resultant powers are determined and illustrated. Good accuracy in comparison with reference values is obtained. The model is applied to two types of reactors. There are modular of fast reactor design like PRISM reactor[7], and modular high temperature gas-cooled reactor design like HTR-M reactor[6]. The PRISM reactor is fuelled by 239Pu, the HTR-M reactor is fuelled by ^{235}U as fissile nuclides. In the work of Van Dam[2], (we used it for comparison purpose), the author obtained reactivity accident due to negative temperature feedback after loss of cooling to different reactors with different fissile material. The reactivity- initiated accident is considered to be due to a linear temperature feedback, and an adiabatic heating of the core after loss of cooling. In the present work, we consider reactivity accident due to a linear temperature feedback, an adiabatic heating of the core after loss of cooling with the addition positive reactivity of the control rods. We analyzed accidents in different types of reactors (HTR-M and PRISM), using the stiffness confinement method for solving the kinetic equations. In the present work, one obtains reactivity induced accident due to control rods ejection of (negative temperature feedback, and addition positive reactivity of the control rods) to overcome the occurrence of the control rods ejection accident, and prevent reactors from damage. The addition positive reactivity of the control rods has four cases: $(0, \beta/2, 0.8\beta, \beta)$, where at the zero case only negative temperature feedback as the case of[2], and the other cases negative temperature feedback with the addition positive reactivity of control rods, this is called reactivity - initiated accident. For ^{239}Pu fueled reactor, when reactivity of the reactor is increased by β, the reactor peak power increases by 83.8085 times of the initial value with saturated temperature of 1,503 (K). For HTR-M reactor, power increases by a factor of 53.5 times the initial value at equilibrium temperature of 1,000 (K) when the reactivity is increased by β.

REFERENCES

1.   Y. Chao, Al. Attard, A resolution to the stiffness problem of reactor kinetics, Nuclear Science and Engineering 90 (1985) 40-46.

2.   H.Vandam, Dynamics of passive reactor shutdown, Prog.Nucl.Energy, 30, (1996), 255.

3.   A. Nahla, E. M. E. Zayed,"Solution of the nonlinear point nuclear reactor kinetics equations", Prog. Nucl. Energy, P. (1-4), (2010).

4.   J.J. Duderstadt, L.J, L.J, L.J. Hamilton, Nuclear Reactor Analysis. John

Wiley & Sons, 1976, pp. 233-251.

5. D. McMahon, A. Pierson, A Taylor series solution of the reactor point kinetics equations, arXiv: 1001.4100, V2 (2010) 1-13.

6. K. Kugeler, R. Schulten, High Temperature Reactor Technology, Springer, Berlin, 1989, pp. 246-260.

7. G.J. Van Tuyle, G.C. Slovik, R.J. Kennett, B.C. Chan, A.L. Aronson, Analyses of unscramed events postulated for the PRISM design, Nucl. Technol.91 (1990) 165-184.

Chapter 10

STUDY OF NATURAL RADIOACTIVITY AND RADIOLOGICAL HAZARD OF SAND, SEDIMENT, AND SOIL SAMPLES FROM INANI BEACH, COX'S BAZAR, BANGLADESH

M. M. Ahmed[1], S. K. Das[1], M. A. Haydar[2], M. M. H. Bhuiyan[2], M. I. Ali2, D. Paul[2]

[1]Department of Physics, Jagannath University, Dhaka, Bangladesh

[2]Health Physics & Radioactive Waste Management Unit, INST, BAEC, Savar, Dhaka, Bangladesh

ABSTRACT

The radionuclide contents and their activity concentrations in beach sand, sediment and adjacent soil samples collected from Inani sea beach, Cox's Bazar, Bangladesh were determined by using a high resolution germanium detector (HPGe) of 20% relative efficiency. A total of eighteen samples of three categories were collected and analysed. The activity concentrations of ^{226}Ra, ^{232}Th and ^{40}K in the sand samples of the beach were found to be varied from 15.14 ± 2.62 to 28.67 ± 3.09 Bqkg^{-1}, 24.39 ± 2.50 to 49.46 ± 3.58 Bqkg^{-1} and 362.00 ± 79.61 to 560.87 ± 81.40 Bqkg^{-1} respectively. For sediment samples the activity concentrations of the corresponding radionuclides ranged between 18.09 ± 2.66 to 53.32 ± 4.01 Bqkg^{-1}, 31.01 ± 2.73 to 78.37 ± 4.35 Bqkg^{-1}, 390.26 ± 76.37 to 733.61 ± 85.80 Bqkg^{-1}, respectively where as for soil samples the values were 26.11 ± 2.99 to 61.66 ± 5.88 Bqkg^{-1}, 41.93 ± 4.18 to 89.39 ± 6.15 Bqkg^{-1} and 467.16 ± 77.62 to 1304.11 ± 147.07 Bqkg^{-1}respectively. The calculated average absorbed dose rates due to these radionuclides in the three types of samples were found as 52.24 nGyh^{-1}, 62.96 nGyh^{-1}, and 104.97 nGyh^{-1}, respectively. The estimated outdoor annual effective dose ranged from 0.05 to 0.08 mSvy^{-1}, 0.05 to 0.11 mSvy^{-1} and 0.07 to 0.17 mSvy^{-1} with the mean values of 0.06 mSvy^{-1}, 0.08 mSvy^{-1}, and 0.13 mSvy^{-1} for sand, sediment, and soil samples, respectively. The values of radium equivalent activity in almost all the samples were less than 370 Bq.kg^{-1}. On the other hand, the values of external hazard index for sand, sediment, and soil samples varied between 0.22 to 0.37, 0.25 to 0.54, and 0.33 to 0.78 with the average values of 0.30, 0.37, and 0.59, respectively. The values are less than unity in all the samples that indicate the non-hazardous nature of the samples.

INTRODUCTION

Natural radioactivity is present in the human environment due to the presence of cosmogenic and primordial radionuclides in the earth's crust [1]. Cosmogenic radionuclides are produced by the interaction of cosmic-rays with atomic nuclei in the atmosphere, while primordial ones (terrestrial background radiation) were formed by the process of nucleo-synthesis. Human activities like mining and milling of mineral ores, ore processing and enrichment etc. cause the redistribution of the activity concentration on the earth crust. Most of the radioactivity deposited on surface sediment is washed by rains and drained through rivers to the oceans. The naturally occurring radionuclides are relatively and uniformly distributed in the seas and oceans. Radionuclides on reaching the oceans become part of the marine ecosystem (water, sediments, sands and biota) and may transfer- through seawater-sediments-biota interface to human beings [2]. Accumulation of such substances in the marine costal environment raises many problems concerning the safety of biotic life, food chain and ultimately human. Due to the fact that radionuclides can have harmful effects on the habitat and can also pose health hazard problems for human, the assessment of gamma radiation dose from natural sources is of particular importance as natural radiation is the largest contributor to the external dose of the world population [3, 4].

Beaches are deposition landforms, and are the result of wave action by which waves or currents move sand or other loose sediments of which the beach is made of as these particles are held in suspension. Beach materials come from erosion of rocks offshore, as well as from headland erosion and slumping producing deposits of scree. These deposits found at different levels within the sand contain natural radionuclides that contribute to ionizing radiation exposure on earth [5].

Inani beach, recently recognized as one of the most popular tourist spots in Bangladesh, is located at a distance of 35 km to the south of Cox's Bazar sea beach, the longest sea beach in the world. The beach composed of light gray to gray sands [6] has become an attractive spot for tourist due to its fine sands and calm and quiet natural environment. The beach sand is fine to medium in size and its gray color indicates the presence of considerable quantity of fine material mixed with the sand. Since the sand, sediment and the soil adjacent to the beach may contain elevated level of natural radioactivity, therefore, the aim of the present study is to investigate the presence of probable natural radionuclides (^{226}Ra, ^{232}Th and ^{40}K) and their activity concentration levels in the collected samples and finally determine the radiological effect on the tourist, public and the environment due to these radionuclides [7].

METHODS AND MATERIALS

Study Area

Inani beach is the area of interest in the present study, which is located in Ukhia Thana under Cox's Bazar district of Chittagong division. The locations of sample collection under current study are shown in Figure 1 and 2.

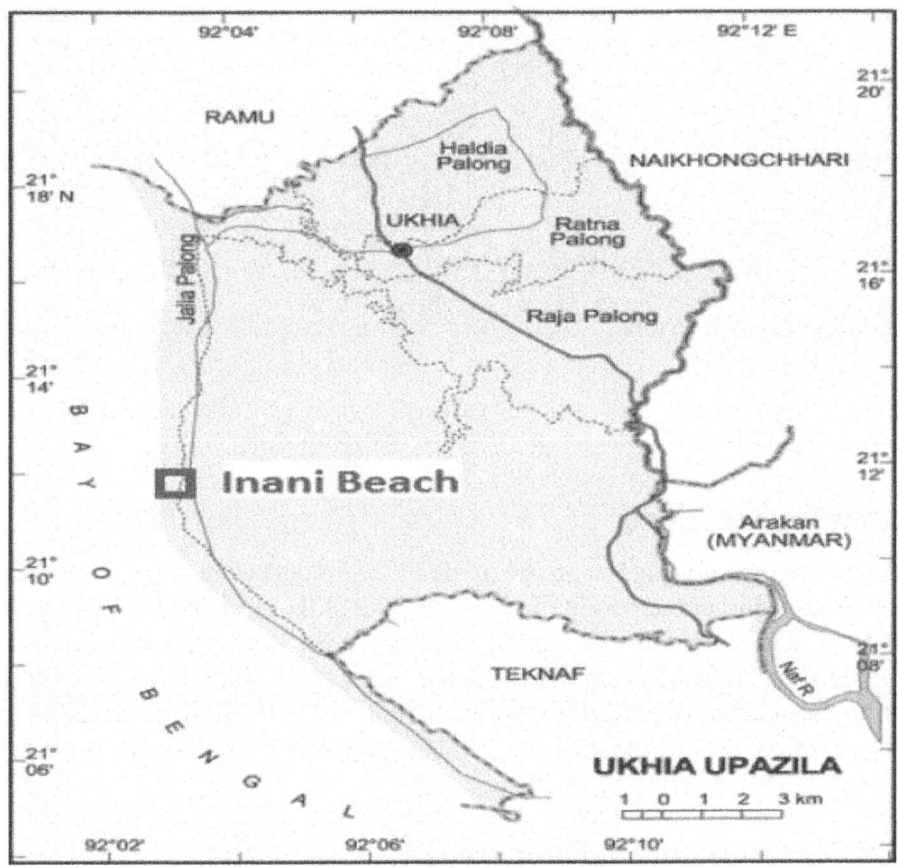

Figure 1: Location map the Inani Beach, Cox's Bazar, Bangladesh

Figure 2: Location map of sampling (using Google map) at the Inani Beach, Cox's Bazar, Bangladesh

Sampling Locations

A total of 18 samples (5-sand, 6-soil and 7-sediment) were collected from the beach area during the period of 11/04/2013 to 13/04/2013. In the present work the soil samples were collected from the adjacent area of the beach, most of which is the cultivated land, whereas sand and sediment samples were collected from the beach area and shore line of the sea, respectively. The coordinates of the sampling area of Inani beach is in between 21°8'55.52"N to 21°9'49.79"N and 92°4'19.11"E to 92°3'21.33"E.

Sample Collection and Processing

About 1 to 1.25 kg of sample was collected from each sampling location. The samples were collected from equidistant locations with a distance of 100 m from each other and at a depth of 10-15 cm from the surface. Upon collection, all the samples were properly packed and marked for their identification code. The samples were then transported, stored and processed at the sample preparation laboratory of Health Physics and Radioactive Waste management Unit (HPRWMU), Bangladesh Atomic Energy Commission (BAEC), Savar, Dhaka. The samples were cleaned for stones, gravels, grass-roots, vegetation etc. and then dried in the sun for several days. The samples were then crushed and ground to fine grains using a grinding machine. The samples were mixed homogeneously and passed through a sieve of 200 μm mesh size. The samples were dried again in a temperature controlled oven at 104°C temperature for 24 hours in order to eliminate any traces of water. Upon drying the samples were transferred to sealable cylindrical plastic containers of 7 cm height and 5.5 cm in diameter and the weights (after grinding) of the samples were recorded using an electrical balance.The sample-filled plastic containers were sealed tightly with cap and wrapped with thick vinyl tape around their necks; marked individually with identification number, sample location, date of preparation and net weight and then stored for about 30 days to assume secular equilibrium between ^{238}U and ^{232}Th series and their daughter progenies.

Experimental Procedure

The detection and measurement of radionuclides in the samples were carried out by gamma spectrometry system using a p-type co-axial HPGe detector of 93 cm^3 active volume and 20% relative efficiency supplied by CANBERRA *(Model GC-2018 and serial No. 0408941)*. The co-axial geometry with electrical contacts in the form of concentric cylinders closed at the end makes it possible to produce very large volume detector elements with excellent efficiencies for high-energy photons.

The HPGe had a resolution of 2 keV at 1332 keV of Cobalt-60 gamma-ray line. The detector was coupled to a 16 k-channel analyser. The spectra of all samples were perfectly analysed using Genie-2000 spectra analysis software (which matched various gamma energy peaks to a library of all possible radionuclides) to calculate the concentrations of ^{226}Ra, ^{232}Th and ^{40}K. The detector was enclosed in a cylindrical shielding container made of lead and iron and having a moving cover to reduce the external gamma-ray background. All the samples were counted for 10 ks. Prior to the measurement of the samples, the environmental gamma background at the laboratory site was determined with an identical empty plastic container used in the sample measurement. The energy regions selected for the corresponding radionuclides were 295 keV and 352 keV of ^{214}Pb and 609 keV, 1120 keV and 1764 keV for ^{214}Bi for ^{226}Ra, 583 keV and 2614 keV of ^{208}Tl, 911 keV and 969 keV of ^{228}Ac for ^{232}Th, and 1460 keV for ^{40}K [8]. These are shown in Figure 3(a).

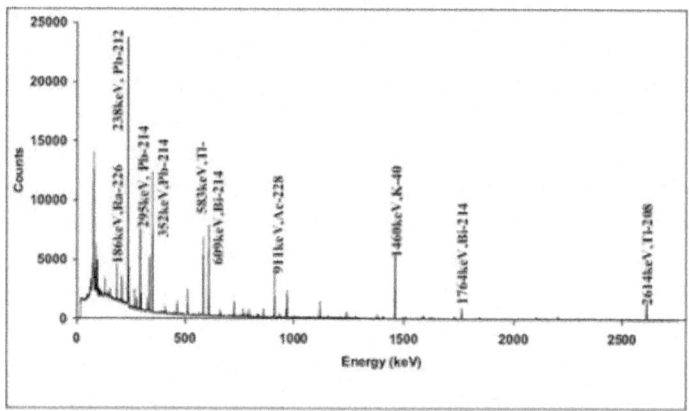

Figure 3(a): Energy diagram of the sample in HPGe detector with 20% efficiency

Calibration of the Detector

The efficiency calibration of the detector was carried out by standard source of solid matrix prepared using ^{226}Ra standard solution. The standard source was prepared using identical container used for the measurement of the samples. The preparation process of standard sources had been reported elsewhere [9]. The efficiency calibration curve as a function of energy for solid matrix is shown in Figure 3(b). The energy calibration of the detector was performed by ^{137}Cs and ^{60}Co point sources.

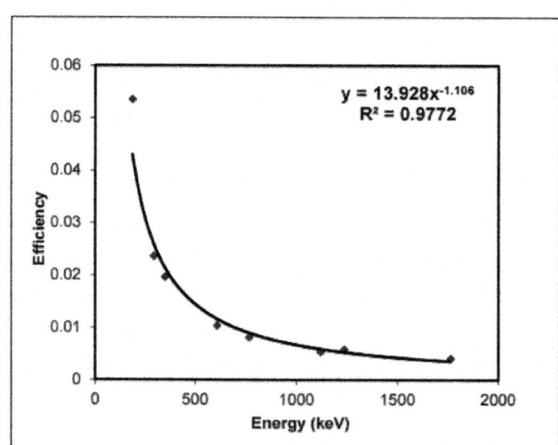

Figure 3(b): Efficiency curve of the HPGe detector for Solid matrix

Calculation of Elemental Concentrations

The activity concentration (A) of each radionuclide in the samples was determined by using the net count (cps) (found by subtracting the background counts from the gross counts with same counting time under the selected photo peaks), weight of the sample, the photo-peak efficiency and the gamma intensity at a specific energy as [10]:

$$A = \frac{cps \times 1000}{E \times I \times W}$$

(1)

Where, A = activity of the sample in Bqkg⁻¹

cps = the net counts per second = cps for the sample − cps for the background value.

E=the counting efficiency of the gamma energy

I= absolute intensity of the gamma ray and

W= samples net weight (in kg).

The errors in the measurement have been expressed in terms of standard deviation ($\pm 2\sigma$), where σ is expressed as [11]:

$$\sigma = \left[\frac{N_s}{T_s^2} + \frac{N_b}{T_b^2} \right]^{\frac{1}{2}}$$

(2)

Where, N_s is the counts measured in time T_s and N_b is the background counts measured in the T_b. The standard deviation ($\pm 2\sigma$), in cps was converted into activity in Bqkg⁻¹ according to equation (1).

Calculation of Radiological Hazard

Radiological impacts of the radionuclides found in the samples were calculated on the basis of calculation of radium equivalent activity, dose rate calculation, and effective dose rate and hazard indexes as following:

The radionuclides ^{226}Ra, ^{232}Th and ^{40}K are not homogeneously distributed in soil, sand and sediments samples. The inhomogeneous distribution from naturally occurring radionuclides is due to disequilibrium between ^{226}Ra and its decay products. For uniformity in exposure estimates, the radionuclide concentrations are defined in terms of 'Radium equivalent activity' (Ra$_{eq}$) in Bqkg⁻¹. This allows comparison of the specific activity of materials containing different amounts of ^{226}Ra, ^{232}Th and ^{40}K according to Beretka and Mathew as follows [12]:

$$Ra_{eq} (Bqkg^{-1}) = C_{Ra} + 1.43\ C_{Th} + 0.077\ C_K \tag{3}$$

Where, C_{Ra}, C_{Th} and C_K are the specific activities of ^{226}Ra, ^{232}Th and ^{40}K, respectively in Bqkg^{-1}.

The absorbed dose rate was calculated from the measured activities of ^{226}Ra, ^{232}Th and ^{40}K in the surface soil, sand, and sediment samples using the formula given below [13]:

$$D\ (nGyh^{-1}) = 0.462C_{Ra}+0.604C_{Th}+0.042C_K \tag{4}$$

Where, D is absorbed dose rate $(nGyh^{-1},)$. In natural environmental radioactivity situations, the effective dose is calculated from the absorbed dose by applying the factor is 0.7 Sv/Gy [14].

To estimate the annual effective dose rate, the conversion co-efficient from absorbed dose, 0.7 SvGy^{-1} and outdoor occupancy factor 0.2 proposed by UNSCEAR, 2000 were used by considering that the people on the average, spent 20% of their time outdoors [15]. The effective dose rate in units of mSvy^{-1} was calculated by the following formula:

$$E\ (mSvy^{-1}) = D \times 365.25 \times 0.2 \times 0.7\ S \times 10^{-6} \tag{5}$$

The external hazard index, H_{ex}, is defined as [16]:

$$H_{ex} = (C_{Ra}/370+C_{Th}/259+C_K/4810)\leq 1 \tag{6}$$

The value of this index must be less than unity in order to keep the radiation hazard insignificant.

RESULTS AND DISCUSSIONS

The results of the present study on the three types (Sand, Sediment and Soil) of samples are summarized in following sections.

Activity Concentrations in Sand, Sediment and Soil Samples

The activity concentrations of ^{226}Ra, ^{232}Th and ^{40}K of all the samples were calculated in Bqkg^{-1} with a counting error of two sigma ($\pm 2\sigma$) and are shown in the Table 1, 2 & 3. The bar diagrams (Figure 4, 5 & 6) show the distribution of the concentrations of all the radionuclides present in the collected samples.

Table 1: Activity Concentration (Bqkg-1) in Different Sand Samples of Inani Beach

Sample code	Activity concentration in Bqkg⁻¹		
	Ra-226	Th-232	K-40
IN SAND 01	15.14 ± 2.62	26.70 ± 2.65	362.00 ± 79.61
IN SAND 02	25.16 ± 3.05	49.46 ± 3.58	534.21 ± 82.70
IN SAND 03	28.67 ± 3.09	48.42 ± 4.32	516.34 ± 78.87
IN SAND 04	22.18 ± 2.87	34.23 ± 2.95	560.87 ± 81.40
IN SAND 05	17.68 ± 2.70	24.39 ± 2.50	414.43 ± 76.81
Average	21.76 ± 2.87	36.64 ± 3.20	477.57 ± 79.88
World AVG	35	30	400

Table 2: Activity Concentration (Bqkg-1) in Different Sediment Samples of Inani Beach

Sample code	Activity concentration in Bqkg⁻¹		
	Ra-226	Th-232	K-40
IN SEDI 01	24.61 ± 2.84	43.60 ± 3.20	475.50 ± 75.66
IN SEDI 02	53.32 ± 4.01	78.37 ± 4.35	450.48 ± 79.60
IN SEDI 03	29.95 ± 3.34	58.85 ± 3.96	412.10 ± 83.69
IN SEDI 04	18.19 ± 2.83	32.86 ± 2.97	481.60 ± 85.80
IN SEDI 05	18.09 ± 2.66	31.01 ± 2.73	390.26 ± 76.37
IN SEDI 06	25.29 ± 3.02	46.73 ± 3.39	733.61 ± 85.80
Average	28.17 ± 3.12	48.57 ± 3.43	490.59 ± 81.04
World AVG	35	30	400

Table 3: Activity Concentration in Bqkg-1 in Different Soil Samples of Inani Beach

Sample code	Activity concentration in Bqkg⁻¹		
	Ra-226	Th-232	K-40
IN SOIL 01	38.86 ± 4.30	64.81 ± 4.65	919.32 ± 115.16
IN SOIL 02	44.05 ± 5.66	66.84 ± 5.68	1229.75 ± 162.46
IN SOIL 03	61.66 ± 5.88	89.39 ± 6.15	1304.11 ± 147.02
IN SOIL 04	49.10 ± 5.54	73.30 ± 5.65	1006.96 ± 146.09
IN SOIL 05	48.86± 5.35	77.37 ± 5.68	1214.30 ± 143.36
IN SOIL 06	42.09 ± 4.68	74.92 ± 6.65	909.09 ± 124.22
IN SOIL 07	26.11 ± 2.99	41.93 ± 4.18	467.16 ± 77.62
Average	44.39 ± 4.91	69.79 ± 5.52	1007.25 ± 130.85
World AVG	35	30	400

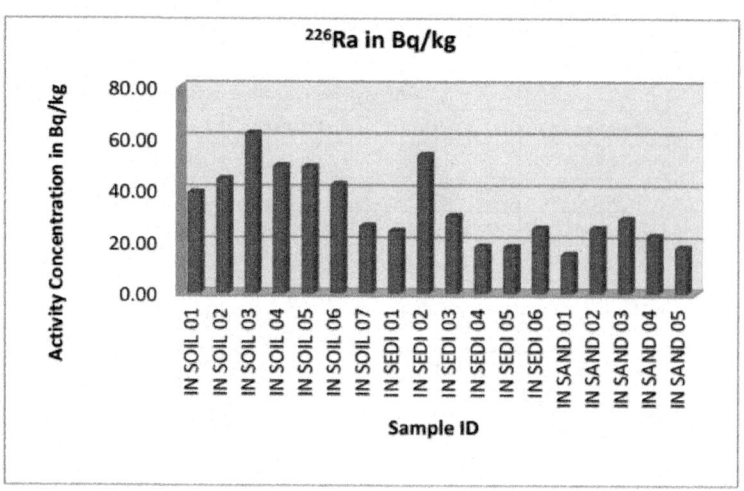

Figure 4: Distribution of ^{226}Ra in all samples of Inani beach, Cox's Bazar, Bangladesh

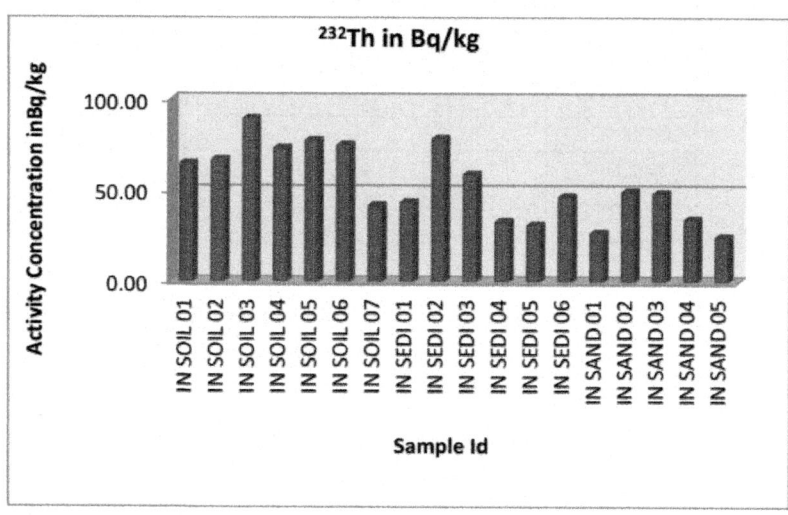

Figure 5: Distribution of ^{232}Th in all samples of Inani beach, Cox's Bazar, Bangladesh

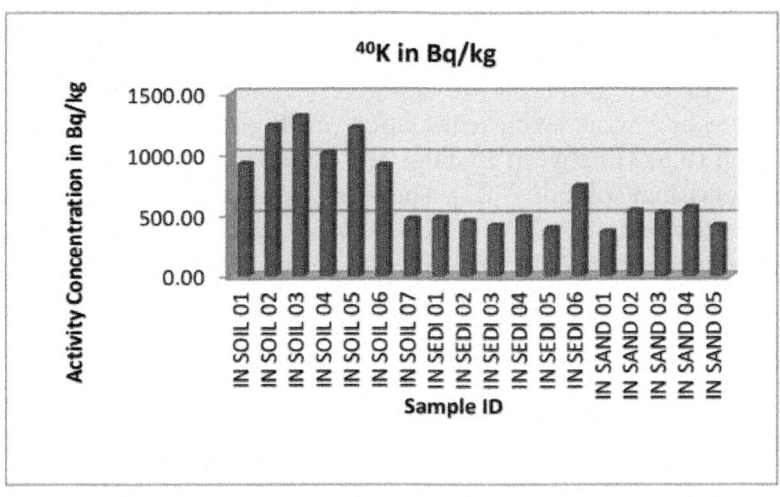

Figure 6: Distribution of ^{40}K in all samples of Inani beach, Cox's Bazar Bangladesh

The activity concentrations of ^{226}Ra, ^{232}Th, and ^{40}K in sand samples were found to be varied from 15.14 ± 2.62 Bqkg^{-1} to 25.16 ± 3.05 Bqkg^{-1}, 24.39 ± 2.50 Bqkg^{-1} to 49.46 ± 3.58 Bqkg^{-1}, and 362.00 ± 79.61 Bqkg^{-1}to 560.87 ± 81.40 Bqkg^{-1} with the average values of 21.26 ± 2.87 Bqkg^{-1}, 36.64 ± 3.20 Bqkg^{-1}, and 477.57 ± 79.80 Bqkg^{-1}, respectively. The average values are almost same as measured by K. M. N. Islam et al at Kuakata Sea-Beach of Bangladesh [17].

The activity concentrations of ^{226}Ra, ^{232}Th, and ^{40}K in sediment samples were ranged from 18.09 ± 2.66 Bq.Kg^{-1} to 53.32 ± 4.01 Bq.Kg^{-1}, 31.01 ± 2.73 to 78.37 ± 4.35 Bq.Kg^{-1} and 390.26 ± 76.37 to 733.61 ± 85.80 Bq.Kg^{-1} with an average of 28.17 ± 3.12 BqKg^{-1}, 48.57 ± 3.43 BqKg^{-1}, and 490.59 ± 81.04 BqKg^{-1}, respectively. These average values are almost similar the values found in another study on the sediment samples collected from a tourist spot in Bangladesh [18].

The activity concentrations of ^{226}Ra, ^{232}Th, and ^{40}K in soil samples ranged from 26.11 ± 2.99 Bqkg^{-1}to 61.66 ± 5.88 Bqkg^{-1}, 41.93 ± 4.18 Bqkg^{-1}to 89.39 ± 6.15 Bqkg^{-1}, and 467.16 ± 77.62 Bqkg^{-1} to 1304.11 ± 147.07 Bqkg^{-1}with the average values of 44.39 ± 4.91 Bqkg^{-1}, 69.79 ± 5.52 Bqkg^{-1}, and 1007.25 ± 130.85 Bqkg^{-1}, respectively. The results found in the current study show that the activity levels in sea-beach soil are almost same as the main land soil as per other report [19].

The correlation of naturally occurring radionuclides has been shown in Figures 7, 8 and 9. ^{226}Ra and ^{232}Th were strongly correlated with correlation coefficient R^2 value of 0.945. This strong correlation might be due to they have some similarity in there environmental origin [23]. A good correlation between [226]Ra and [232]Th is shown in Figure. 7. The relationship between [226]Ra and [40]K, were weak with correlation coefficient of 0.613. The weak correlation (0.613) between Radium and potassium might be explained due to high potassium solubility [23]. The weak correlation existed between [232]Th and [40]K in the samples ((R^2 = 0.585)shown in Figure. 9, which indicated [40]K concentrations might not be related to the presence of [232]Th [24].

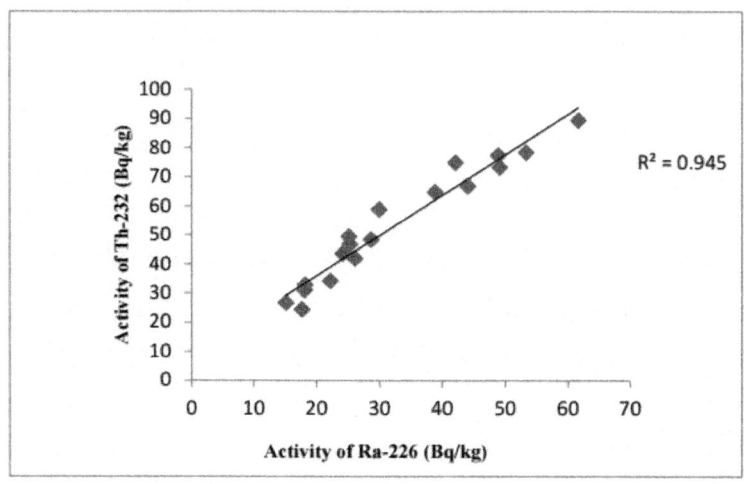

Figure 7: Correlation between activity concentrations of [226]Ra and [232]Th in all samples

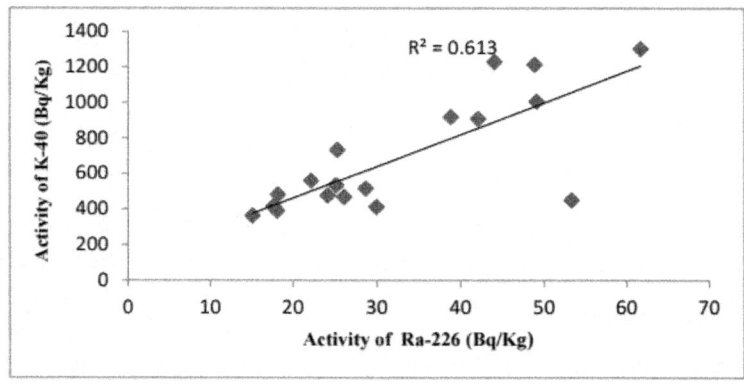

Figure 8: Correlation between activity concentrations of [226]Ra and [40]K in all samples

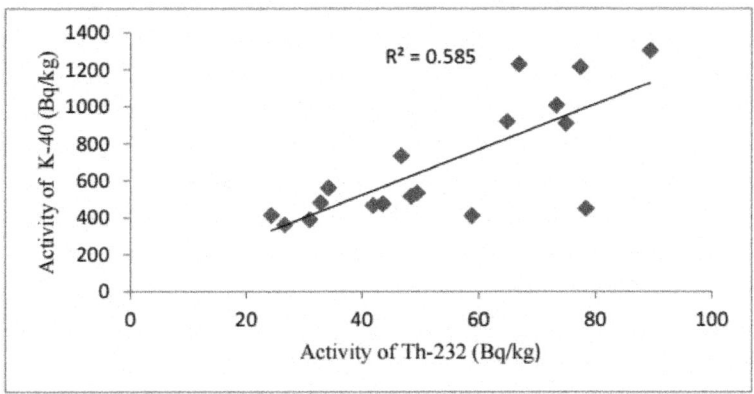

Figure 9: Correlation between activity concentrations of ^{232}Th and ^{40}K in all samples

Moreover, the values of activity concentrations of ^{226}Ra, ^{232}Th, and ^{40}K found in the current study for few samples were higher than the worldwide average value of 35, 30 and 400 Bqkg^{-1}, successively [20].

Radiological Indices

The radiological parameters such as indices of radium equivalent activity (Ra$_{eq}$), absorbed dose rate (D), outdoor annual effective dose (E), and external hazard index (H$_{ex}$) were calculated to estimate the radiological risk due to the presence of ^{226}Ra, ^{232}Th and ^{40}K in the samples. Table 4 depicts the values of Ra$_{eq}$, D, E, and H$_{ex}$.

The values of radium equivalent activity sand, sediment, and soil were found to be varied from 81.12 Bqkg^{-1} to 137.56 Bqkg^{-1}, 92.42 Bqkg^{-1} to 199.93 Bqkg^{-1} and 121.95 Bqkg^{-1} to 289.68 Bqkg^{-1} with an average of 110.84 Bqkg^{-1}, 135.29 Bqkg^{-1}, and 221.58 Bqkg^{-1}, respectively which were lower than the world average value of 370 Bqkg^{-1} [21] reported by OECD.

The calculated values of external hazard index for sand, sediment and soil samples ranged from 0.22 to 0.37, 0.25 to 0.54, and 0.33 to 0.78 with an average of 0.30, 0.37, and 0.59, respectively which were lower than the world wide average value 1 [20].

The absorbed dose rate due to the terrestrial gamma rays at 1m above from the ground were in the range of 40.30 to 64.18 nGyh^{-1}, 43.48 to 90.89 nGyh^{-1} and 57.01 to 137.25 nGyh^{-1} with the average values of 52.24 nGyh^{-1}, 62.96 nGyh^{-1} and 104.97 nGyh^{-1}, respectively and the values (Sediment and Soil Samples) are higher than the world average value of 55 nGyh^{-1} [22].

Table 4: Absorbed dose rate (D), outdoor annual effective dose (E), radium equivalent activity (Ra$_{eq}$) and external hazard index (H$_{ex}$) of the samples collected from Inani Beach Cox's Bazar

Sample ID	Radium equivalent activity, Ra$_{eq}$ (Bqkg^{-1})	Hazard Index, H$_{ex}$	Absorbed dose rate, D (nGyh^{-1})	Annual effective dose, E (mSvyr^{-1})
INSAND 01	81.12	0.22	38.32	0.05
INSAND 02	136.90	0.37	63.93	0.08
INSAND 03	137.56	0.37	64.18	0.08
INSAND 04	114.23	0.31	54.48	0.07
INSAND 05	84.40	0.23	40.30	0.05
AVG	**110.84**	**0.30**	**52.24**	**0.06**
IN SEDI 01	123.03	0.33	57.47	0.07
IN SEDI 02	199.93	0.54	90.89	0.11
IN SEDI 03	145.72	0.39	66.69	0.08
IN SEDI 04	102.18	0.28	48.48	0.06
IN SEDI 05	92.42	0.25	43.48	0.05
IN SEDI 06	148.48	0.40	70.72	0.09
AVG	**135.29**	**0.37**	**62.96**	**0.08**
IN SOIL 01	202.17	0.55	95.71	0.12
IN SOIL 02	234.13	0.63	112.37	0.14
IN SOIL 03	289.68	0.78	137.25	0.17
IN SOIL 04	231.27	0.63	109.25	0.13
IN SOIL 05	252.80	0.68	120.31	0.15
IN SOIL 06	219.05	0.59	102.88	0.13
IN SOIL 07	121.95	0.33	57.01	0.07
AVG	**221.58**	**0.59**	**104.97**	**0.13**
World AVG	**370**	...	**55**	**0.07**

The outdoor annual effective dose equivalent was calculated from the air absorbed doses using the relation given in equation (5). The values varied from 0.05 to 0.08 mSvy^{-1}, 0.05 to 0.11 mSvy^{-1} and 0.07 to 0.17 mSvy^{-1} with the mean values of 0.06 mSvy^{-1}, 0.08 mSvy^{-1} and 0.13 mSvy^{-1} for sand, sediment, and soil samples, respectively. The values in two types of samples are just higher than the worldwide average values for outdoor annual effective dose of 0.07 mSvy^{-1} [20].

The values of hazard index are lower than the unity, absorbed dose rate and annual effective dose found (Sediment and Soil samples) in the present study just exceed the standard limits for radiological safety. Therefore, though insignificant, these environmental components are unable to produce higher radiation exposure effect to the public living around the area. Therefore, no harmful radiation effects are posed to the public and tourists going to the beaches for recreation.

CONCLUSIONS

The activity concentrations and radiological hazard associated with beach sand, sediment and adjacent soil samples collected from the Inani Sea-Be ach, Cox's Bazar of Bangladesh were investigated in the present study. The results indicated that only the natural radionuclides were present in the samples. The radioactivity concentrations of ^{226}Ra, ^{232}Th and ^{40}K were just similar to the world average values. The activity levels in sand and sediment samples were also comparable to the other parts of the country. The calculated average absorbed dose rate and estimated outdoor annual effective dose were found lower than the worldwide average values. The average radium equivalent activity (R_{eq}) and external hazard (H_{ex}) index were less than the world average value. Therefore, the probability of the radiological impact on the inhabitants/ public living in this area will be insignificant and there is no probability of radiological health effect on the tourists.

REFERENCES

1. Mirjana B, Radenković, Saeed Masud Alshikh, Veibor B. Andrić and Šćepan S. Miljanić., Radioactivity of sand from several renowned public beaches and assessment of the corresponding environment risks Journal of Serbian Chemical Society. JSCS.74 (4)461-470(2009) JSCS-3847, doi:10.2298/JSC0904461R.

2. Turner, J.E., Atoms, Radiation and Radiation Protection, J. Wiley, New York, 1-13(1995).

3. UNSCEAR, Report of the General Assembly with ScientificAnnexes. http://www.unscear.org/docs/reports/annexb.pdf(2000).

4. Pfennig, H., Klewe-Nebenius, H., Seelmann-Eggebert, W., Karlsruher Nuklidkarte, 6. Auflage (1995).

5. Peterson, J.M., Mac Donell, M., Haroun, L., Monette, F., Hildebrand, R.D., Taboas, A., Radiological and Chemical Fact Sheets to Support Health Risk Analyses for Contaminated Areas, Human Health Fact Sheet, Argonic, 38-39(2007).

6. M. N. Alam, M. I. Chowdhury, M. Kamal, S. Ghose, M. N. Islam and M. N. Mustafa, The 226Ra, 232Th and 40K Activities in Beach Sand Minerals and Beach Soils of Cox's Bazar, Bangladesh, Journal Environmental Radioactivity,   Vol.46,No.2,1999,pp.243-250.doi:10.1016/S0265-

931X(98)00143-X.

7. Kannan, V., Rajan, M.P., Lyengar, M.A. and Ramesh, R. Distribution of natural and anthropogenic radionuclides in soil and beach sand Samples of Kalpakkam (India). Appl. Rodiat. Isot. 57,109-119 (2002).

8. C.E. Roessier, Z.A. Smith, W.E. Bloch and R.J Prince, Uranium and radium in Floride phosphate materials, Health Phys. vol. 37, 1970, pp. 269-277.

9. S. Harb, K.S. Din and A. Abbady, Study of efficiency calibrations of HPGe detectors for radioactivity measurement of environmental samples, Proceedings of the 3rd Environmental Physics Conference, 19-23 Feb. 2008, Aswan, Egypt.

10. International Atomic Energy Agency, Measurement of Radionuclide in Food and the Environment: A Guide- book, International Atomic Energy Agency, Vienna, 1989, pp. 139-144. doi:10.4236/jep.2012.39126 published Online September 2012 (http://www.SciRP.org/journal/jep).

11. G. F. Knoll, Radiation Detection and Measurement, Third edn.1998, New York.

12. Beretka, J. and Mathew, P.J., Natural radioactivity of Australian buildings, materials, industrial wastes and by products, Health Physic, 48, 87-95 (1985).

13. UNSCEAR, 2000. Sources and Effects of Ionizing Radiation. United Nations Scientific Committee on the Effects of Atomic Radiation, United Nations, New York.

14. UNSCEAR, 1993. Sources and Effects of Ionizing Radiation. United Nations Scientific Committee on the Effects of Atomic Radiation, United Nations, New York.

15. K. Debertin and R.G. Helmer. Gamma and X-ray spectrometry detectors, North Holland, 1988.

16. Lu, X. And Xiolan, Z. (2006): Measurement of natural radioactivity in sand samples collected from the Booje Weithe sand park, China, Environ. Geol. 50, 977-988.

17. Khondaker Mohammed Nazrul Islam, Debasish Paul, Md. Mahbubur Rahman Bhuiyan, Amina Akter, Budrun Neher, Sheikh Mohammad Azharul Islam. Study of Environmental Radiation on Sand and Soil Samples from Kuakata Sea Beach of Patuakhali. Journal of Environmental Protection, 2012, 3, 1078-1084.

18. Alamgir Miah, M. M. H. Miah, Masud Kamal, M. I. Chowdhury, M. Rahmatullah; Natural Radioactivity and Associated Dose Rates in Soil

Samples of Malnichera Tea Garden Sylhet District of Bangladesh, Journal of Nuclear and Particle Physics 2012, 2(6): 147-152, DOI: 10.5923/j. jnpp.20120206.03.

19. M. I. Chowdhury, M. Kamal, M. N. Alam, Saleha Yeasmin and M. N. Mostafa, Distribution of Naturally Occurring Radionuclides in Soils of Southern Districts of Bangladesh. Radiation Detection Dosimetry (2005), 1of 5.doi:10.1093/rpd/nci335.

20. UNSCEAR, Sources and effects of ionizing radiation, Report of the United Nations Scientific Committee on the Effects of Atomic Radiation to the General Assembly, United Nations, New York, USA, (2000). Annex.

21. United Nations Scientific Committee on the Effects of Atomic Radiation (UNSCEAR), Sources, Effects and Risks of Ionising Radiation, United Nations, New York, 1988.

22. M. R. Abdil, H. Faghihian, M. Kamali, M. Mostajbod-davati and A. Hasanzadeh, Distribution of Natural Radionuclides on Coasts of Bushehr, Persian Gulf, Iran, Iranian Journal of Science & Technology: Transaction A, Vol. 30, No. A3, 2006, pp. 259-269.

23. M. H El Mamoney, Ashraf E. M. Khater, Environmental Characterization and Radio-ecological Impacts of Non-nuclear Industries on the Red Sea Coast. Journal of Environmental Radioactivity 73(2004) 151-168.

24. S. Harb, Natural Radioactivity and External Gamma Radiation Exposure at The Costal Red Sea in Egypt. Radiation Protection Dosimetry (2008), Vol 130, No 3, pp. 376-384.

Chapter 11

MATHEMATICAL MODELING FOR PULSATILE FLOW OF A NON-NEWTONIAN FLUID WITH HEAT AND MASS TRANSFER IN A POROUS MEDIUM BETWEEN TWO PERMEABLE PARALLEL PLATES

Mohamed Y. Abou-zeid[1, 2], Seham S. El-zahrani[1],
Hesham M. Mansour[3]

[1]Department of Mathematics, Faculty of Science, Tabuk University, Tabuk, KSA

[2]Department of Mathematics, Faculty of Education, Ain Shams University, Roxy, Cairo, Egypt

[3]Department of Physics, Faculty of Science, Cairo University Giza, Egypt

ABSTRACT

The aim of the present work is to investigate the effect of mixed convection heat and mass transfer on pulsatile flow of a non-Newtonian fluid which is obeying the rheological equation of state due to Ree-Eyring's stress-strain relation. We take into consideration the porosity of medium and a uniform magnetic field. The flow is assumed to be between two permeable vertical plates. The equations of momentum, energy and concentration have been solved by using Lightill method. The velocity, temperature and concentration distributions are obtained. The effects of various parameters of the problem on these distributions are presented and discussed.

INTRODUCTION

The study of pulsating flows is of practical engineering importance. High speed (turbulent) pulsating flows occur in turbo machinery, rotor blade aerodynamics, reciprocating piston-driven flows, etc. Numerous experimental investigations were focused on fundamental studies of fully developed periodic pipe flows with sinusoidal varying pressure gradients (or flow rates). Low speed (laminar) pulsating flows were studied in order to analyze the flows through small pipes or in the blood circulation systems. Laminar flows are relatively simple for analytical (or numerical) analysis and are a natural choice to provide basic studies of fundamental hydrodynamic effects in pulsating flows [1]. Pulsatile flow has also recently found renewed significance in its application to MEMS microfluidic engineering applications [2]. A complete treatment of the fluid dynamics of steady and pulsatory flow with emphasis on basic mechanics, physics and applications can be seen in Ref. [3].

Several investigations on flow through stenosed artery have been carried out to evaluate the flow characteristics under steady and pulsatile flow conditions. Young and Tsai [4-5] studied the steady and unsteady flows across a stenosis experimentally (See also Siouffi et al. [6]). In fact, blood is a complex rheological mixture showing several non-Newtonian properties, shear-thinning, yield stress, stress relaxation etc. The rheological properties of fluid have important influences on wall shear stress, oscillatory shear index etc. So it is very important to address the significance of non-Newtonian models for the purpose of reliable hemodynamic modeling [7]. Blood may be considered as a Newtonian fluid for the flow within the heart and the aorta of the human cardiovascular system. For blood flow in smaller arteries of diameter 0.5mm, a simple rescaling of the Newtonian viscosity is sufficient to take account of non-Newtonian behavior of the blood. In particular situations blood may behave as a non-Newtonian fluid, even in large arteries, as reported in Nakamura and Sawada [8]. Under diseased conditions, blood exhibits remarkable non-Newtonian properties. Eldabe et al. [9-10] studied pulsatile magneto hydrodynamic viscoelastic flow through a channel bounded by two permeable parallel plates and the effect of couple stresses on pulsatile hydro magnetic poiseuille flow.

The flow through porous media has attracted considerable research activity in recent years because of its important applications notably in the flow of oil through porous rock, the extraction of energy from the geothermal regions, the evaluation of the capability of heat-removal from particulate nuclear fuel debris that may result from a hypothetical accident in a nuclear reactor, the filteration of solids from liquids, flow of ion-exchange beds, drug permeation through human skin, chemical reactors for economical separation or purification of mixtures and so on [11]. Flow through porous media is very prevalent in nature, and therefore the study of flow through a porous medium has become of principle interest in many engineering applications. Thermal and solutal transport by fluid flowing through a porous matrix is a phenomenon of great interest from the theory and application point of view. Heat transfer in the case of homogenous fluid-saturated porous media has been studied in relation to different applications like dynamic of hot underground springs, terrestrial heat flow through aquifer, hot fluid and ignition front displacements in reservoir engineering, heat exchange between soil and atmosphere and heat exchanges with fluidized beds. Mass transfer in isothermal condition has been studied with applications to problems of mixing of fresh and salt water in a quifers, spreading of solutes in fluidized beds and crystal washers, salt leaching in soils, etc. Prevention of salt dissolution into the lake waters near the sea shores has become a serious and interesting problem of research [12].

Verma et al. [13] studied the pulsatile blood flow of a micro deformable fluid. A mathematical model for the study of blood flow through a channel with permeable walls of finite width is discussed by Mishra and Ghosh [14]. Vajravelu et al. [15] studied the pulsatile flow of a viscous incompressible Newtonian fluid between permeable beds. Lalit and Narayanan [16] discussed the analysis of pulsatile flow and its role on particle removal from surfaces. The pulsatile flow of blood through a mild stenosed artery under periodic body acceleration is investigated in this theoretical analysis by Mallik et. al. [17]. The problem of pulsatile flow of MHD non-Newtonian fluid obeying power law model with convective heat transfer through a non-Darcy porous medium between two coaxial cylinders is studied by Abou-zeid [18]. Shawky [19] studied the flow due to the pulsatile pressure gradient of dusty non-Newtonian fluid obeying Eyring-Powell model with heat transfer in a channel in the presence of external magnetic field.

The main idea of the present work is to study the mathematical analysis of a non-Newtonian fluid flow with heat and mass transfer. The flow analysis is developed through a porous media and between two vertical permeable parallel plates, in the presence of a normal magnetic field. Analytical solutions for the momentum, energy and concentration equations have been obtained under the assumption that the pulsatile flow is a linear combination of the steady part and the oscillatory part. Also, we show the relation between the different parameters of the flow and the external forces. Numerical solutions are obtained for different values of The parameters Re , k_0 , Ps , P_0 , Da , M , G_T , G_C , Pr , Sr , Sc , ω and t.

MATHEMATICAL ANALYSIS

We consider the unsteady flow with heat and mass transfer of a viscous, incompressible, and electrically conducting non-Newtonian fluid (biviscosity fluid) in a porous medium between two permeable vertical parallel plates situated at $y = 0$ and $y = h$, under the action of the fluid gradient. The coordinates system used is shown in Fig. 1. The x- axis is taken in the direction of the flow and the y- axis is taken normal to the plates. We assume that a uniform magnetic field B_0 is acting along the y- axis. The fluid is being injected into the wall through $y = 0$ and is being sucked through $y = h$ with uniform velocity V_0.

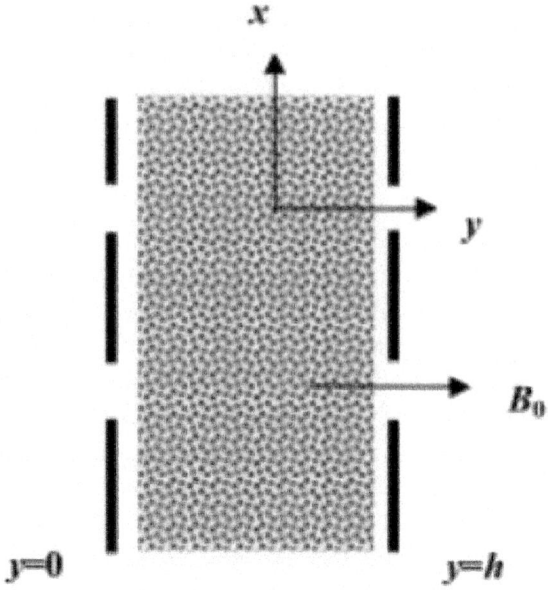

Figure 1: The coordinate system used

The governing equations used in this problem can be written as follows:

Continuity equation

$$\frac{d\rho}{dt} + \rho V_{i,i} = 0 \, ,$$

(1)

Momentum equation

$$\rho \left(\frac{\partial V}{\partial t} + \underline{V} . \nabla \underline{V} \right) = -\nabla P + \nabla \cdot \underline{\tau} - \frac{\mu_f}{k_p} \underline{V} + \underline{J} \times \underline{B} + \underline{F}$$

(2)

Temperature equation

$$\frac{dT}{dt} = k_T \nabla^2 T \, ,$$

(3)

Concentration equation

$$\frac{dC}{dt} = D_m \nabla^2 C + \frac{D_m K_T}{T_m} \nabla^2 T \, ,$$

(4)

Nomenclature

a and b	Characteristics of Eyring-Powell model	R_e	Reynolds number $= \dfrac{h V_0}{\nu_f}$
$\underline{B}$	Total magnetic field	S_c	Schmidt number $= \dfrac{h V_0}{D_m}$
B_0	Uniform magnetic field		
C	Concentration of the species	S_r	Soret number $= \dfrac{D_m k_T (T_1 - T_2)}{h V_0 T_m (C_1 - C_2)}$
C_E	Ergun constant	t	Time
C_p	Specific heat at constant pressure	T	Temperature of the fluid
C_s	Concentration susceptibility	T_m	The mean temperature
Da	Darcy number $= \dfrac{V_0 k_p}{\nu_f h}$	V_0	the velocity of the suction or injection at the walls
D	Coefficient of mass diffusivity	$\underline{V}$	Velocity vector $= (u, v, 0)$
$\underline{F}$	The external body force		
G_T	Temperature Grashof number $= \dfrac{h g \beta^* (T - T_2)}{V_0^2}$		**Greek symbols**
G_C	Concentration Grashof number $= \dfrac{h g \beta (C - C_2)}{V_0^2}$	β	The temperature coefficient of volumetric expansion
h	The distance between the two plates.	β^*	The mass coefficient of volumetric expansion
ι	$\sqrt{-1}$	μ_f	Viscosity coefficient
$\underline{J}$	Current density	ω	Frequency of the oscillating plate
K_T	Thermal conductivity $= \dfrac{k_c}{\rho C_p}$	ρ	Density of the fluid
k_p	Constant permeability	σ	The electric conductivity
k_0	The parameter of the non-Newtonian fluid $= \dfrac{1}{ab\mu_f}$	$\underline{\tau}$	Stress tensor of the fluid.
M	Magnetic parameter $= \dfrac{\sigma B_0^2 h}{\rho V_0}$		**Superscripts and subscripts**
P	Fluid pressure	1	Plate condition at y=0
P_0	The amplitude of the pulsation	2	Plate condition at y=h
P_s	The steady component of the pressure gradient	s	Steady component
		o	Oscillatory component
P_r	Prandtl number $= \dfrac{\nu_f}{k_T}$		

The rheological equation of state for an isotropic and incompressible flow of a Ree-Eyring fluid can be written as

$$\tau_{ij} = \mu_f \frac{\partial V_i}{\partial x_j} + \frac{1}{a}\sinh^{-1}\left(\frac{1}{b}\frac{\partial V_i}{\partial x_j}\right)$$

(5)

Since $\sinh^{-1} x \approx x$ for $|x| \leq 1$, then

$$\tau_{ij} = \mu_f(1+k_0)\frac{\partial V_i}{\partial x_j},$$

(6)

For ordinary Newtonian fluid $k_0 = 0$.

Since the two walls are infinite, all quantities are functions of y and t only. From equation (1), we get $v = V_0$. Equations (2), (3) and (4) reduce to

$$\frac{\partial u}{\partial t} + V_0\frac{\partial u}{\partial y} + \frac{1}{\rho}\frac{\partial P}{\partial x} = v_f(1+k_0)\frac{\partial^2 u}{\partial y^2} - \left(\frac{\sigma B_0^2}{\rho} - \frac{v_f}{k_p}\right)u - g\,\beta(T-T_2) - g\,\beta^*(C-C_2),$$

(7)

$$\frac{\partial T}{\partial t} + V_0\frac{\partial T}{\partial y} = k_T\frac{\partial^2 T}{\partial y^2},$$

(8)

$$\frac{\partial C}{\partial t} + V_0\frac{\partial C}{\partial y} = D_m\frac{\partial^2 C}{\partial y^2} + D_m\frac{k_T}{T_m}\frac{\partial^2 T}{\partial y^2}.$$

(9)

The appropriate boundary conditions are

$$\left.\begin{array}{llllll} u = 0, & T = T_1 & \text{and} & C = C_1 & \text{at} & y = 0, \\ u = 0, & T = T_2 & \text{and} & C = C_2 & \text{at} & y = h, \end{array}\right\}$$

(10)

Let us introduce the following dimensionless quantities as follows:

$$\left.\begin{array}{l} u^* = \dfrac{u}{V_0}, \quad x^* = \dfrac{1}{h}x, \quad y^* = \dfrac{1}{h}y, \quad \omega^* = \dfrac{h}{V_0}\omega, \quad t^* = \dfrac{V_0}{h}t, \\[3mm] P^* = \dfrac{1}{\rho V_0^2}P, \quad T^* = \dfrac{T-T_2}{T_1-T_2}, \quad C^* = \dfrac{C-C_2}{C_1-C_2}, \end{array}\right\}$$

(11)

Hence, Equations (7), (8) and (9) may be written in dimensionless form after dropping the star mark.

$$\frac{\partial u}{\partial t} + \frac{\partial u}{\partial y} + \frac{\partial P}{\partial x} = \frac{1}{Re}(1+k_0)\frac{\partial^2 u}{\partial y^2} - \left(M + \frac{1}{Da}\right)u - G_T\,T - G_C\,C,$$

'(12)

$$\frac{\partial T}{\partial t} + \frac{\partial T}{\partial y} = \frac{1}{Re\, P_r} \frac{\partial^2 T}{\partial y^2},$$

(13)

$$\frac{\partial C}{\partial t} + \frac{\partial C}{\partial y} = \frac{1}{S_c} \frac{\partial^2 C}{\partial y^2} + S_r \frac{\partial^2 T}{\partial y^2},$$

(14)

METHOD OF SOLUTION

For pulsation pressure gradient, let

$$-\frac{\partial P}{\partial x} = \left(\frac{\partial P}{\partial x}\right)_s + \left(\frac{\partial P}{\partial x}\right)_0 e^{i\omega t},$$

(15)

Using Lightill method [20], the system of partial differential equations can be transformed into ordinary differential equations.

$$\left. \begin{array}{l} u = u_s + u_0 e^{i\omega t}, \\[2mm] T = T_s + T_0 e^{i\omega t}, \\[2mm] C = C_s + C_0 e^{i\omega t} \end{array} \right\}.$$

(16)

Substituting from Eq's (16) and (17) in Eq's (12), (13) and (14) and equating the like terms on both sides, we get the following system equations:

$$\frac{1}{Re}(1+k_0)\frac{d^2 u_s}{dy^2} - \frac{du_s}{dy} - \left(M + \frac{1}{Da}\right)u_s - G_r T_s - G_c C_s = -\left(\frac{\partial P}{\partial x}\right)_s,$$

(17)

$$\frac{1}{Re}(1+k_0)\frac{d^2 u_0}{dy^2} - \frac{du_0}{dy} - \left(M + \frac{1}{Da} + i\omega\right)u_0 - G_r T_0 - G_c C_0 = -\left(\frac{\partial P}{\partial x}\right)_0,$$

(18)

$$\frac{1}{Re\, P_r} \frac{d^2 T_s}{dy^2} - \frac{dT_s}{dy} = 0,$$

(19)

$$\frac{1}{R_e P_r} \frac{d^2 T_0}{dy^2} - \frac{dT_0}{dy} - i\omega T_0 = 0,$$

(20)

$$\frac{1}{S_c} \frac{d^2 C_s}{dy^2} + S_r \frac{d^2 T_s}{dy^2} - \frac{dC_s}{dy} = 0,$$

(21)

The dimensionless boundary conditions are

$$u_s = 0, \ u_0 = 0, \ T_s = 1, \ T_0 = 0, \ C_s = 1, \ C_0 = 0 \quad \text{at} \ y = 0 \Big]$$
$$u_s = 0, \ u_0 = 0, \ T_s = T_0 = 0, \ C_s = C_0 = 0 \qquad \text{at} \ y = 1 \Big\}$$

$$(22)$$

The solutions of the equation $(17) \rightarrow (22)$ with boundary conditions (23) are

$$u = a_{23} e^{\alpha_{32} y} + a_{25} e^{\alpha_{33} y} + a_{26} e^{\alpha_{34} y} + a_{27} e^{\alpha_{35} y} + a_{28} e^{\alpha_{36} y} + a_{29} e^{\alpha_{37} y}$$
$$+ a_{41} e^{\alpha_{30} y} + a_{42} e^{\alpha_{24} y} + \left(a_{43} + a_{44} e^{\alpha_{12} y} + a_{45} e^{\alpha_{10} y} \right) e^{i\omega t} ,$$

$$(23)$$

$$T = a_{13} e^{-\alpha_3 y} + a_{14} ,$$

$$(24)$$

$$C = a_{16} e^{-\alpha_3 y} + a_{17} e^{-\alpha_4 y} + a_{18} ,$$

$$(25)$$

where are defined in the appendix.

NUMERICAL RESULTS AND DISCUSSION

The systems of equations that govern the non-Newtonian fluid between two vertical parallel plates are solved analytically. The formulas for velocity, temperature and concentration distributions are obtained. In order to get a physical understanding of the problem and for the purpose of discussing the results, numerical calculations have been performed to obtain the velocity, temperature and concentration. The velocity , temperature and concentration distributions are calculated for different values of Re , k_0, P_s, P_0, Da , M , G_T, G_C , Pr , Sr , Sc , ω and t in figures 2 - 20.

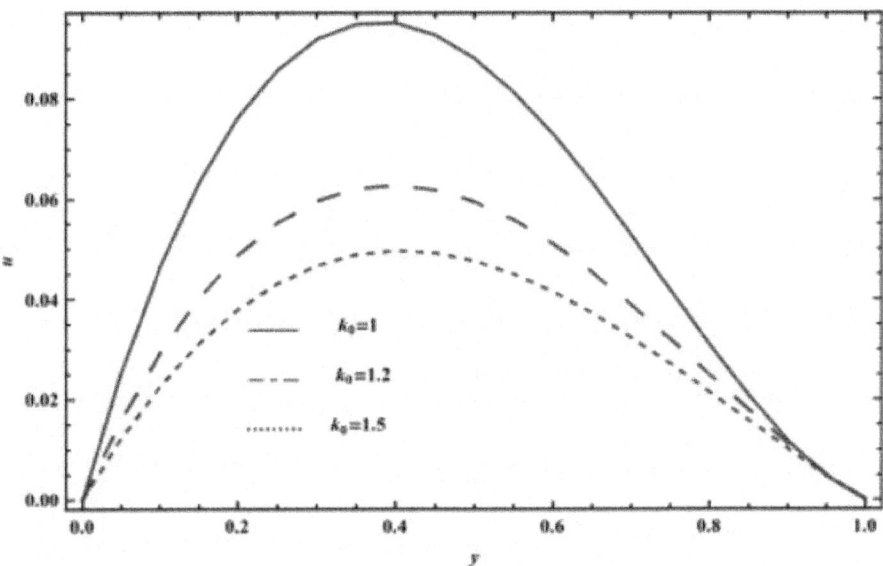

Figure 2: The velocity distribution is plotted versus y, for different values of k_0 and for a system which has the particular values Re =1.6, Da=0.5, M=3, Ps=2.5, P_0=0.5, $\omega=\pi/4$, Pr=1, G_T=3.5, G_C=1.5, Sc=1.5, Sr=1, t=0.5

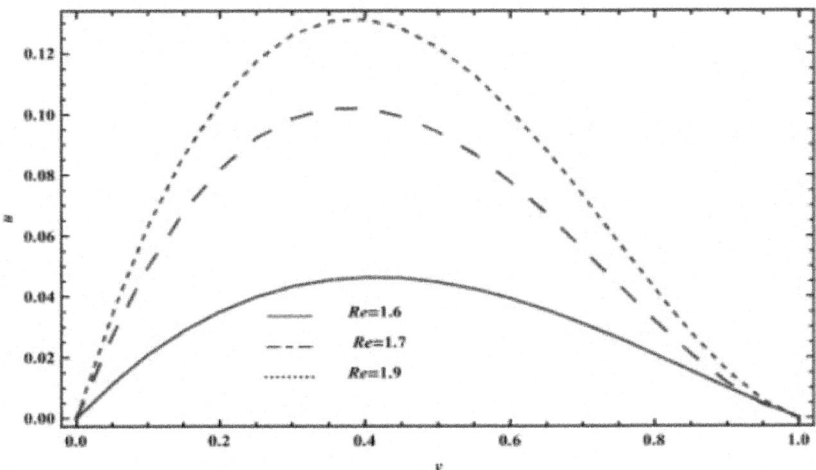

Figure 3: The velocity distribution is plotted versus y, for different values of Re and for a system which has the particular values k_0=1.5, Da=0.5, M=3, Ps=2.5, P_0=0.5, $\omega=\pi/4$, Pr=1, G_T=3.5, G_C=1.5, Sc=1.5, Sr=1, t=0.5

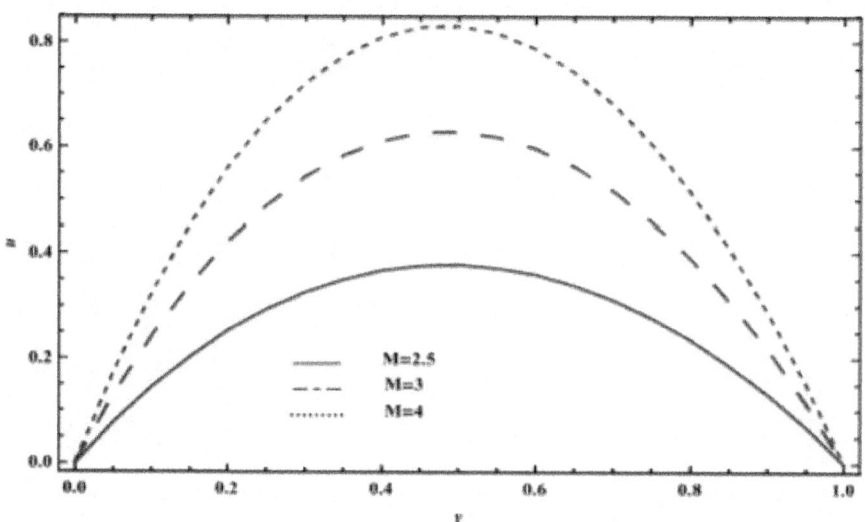

Figure 4: The velocity distribution is plotted versus y, for different values of M and for a system which has the particular values $k_0=1.5$, Re $=1.6$, Da=0.5, Ps=2.5, $P_0=0.5$, $\omega=\pi/4$, Pr=1, $G_T=3.5$, $G_C=1.5$, Sc=1.5, Sr=1, t=0.5

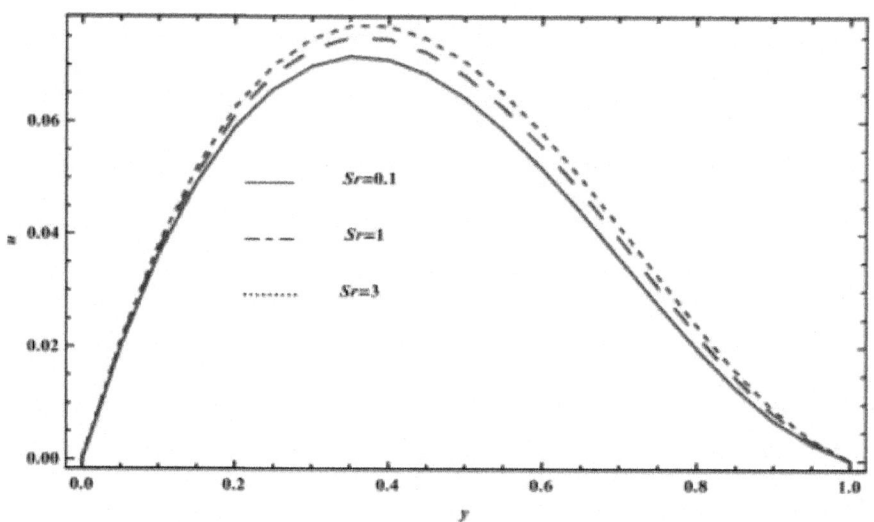

Figure 5: The velocity distribution is plotted versus y, for different values of Sr and for a system which has the particular values $k_0=1.5$, Re $=1.6$, Da=0.5, M=3, Ps=2.5, $P_0=0.5$, $\omega=\pi/4$, Pr=1, $G_T=3.5$, $G_C=1.5$, Sc=1.5, t=0.5

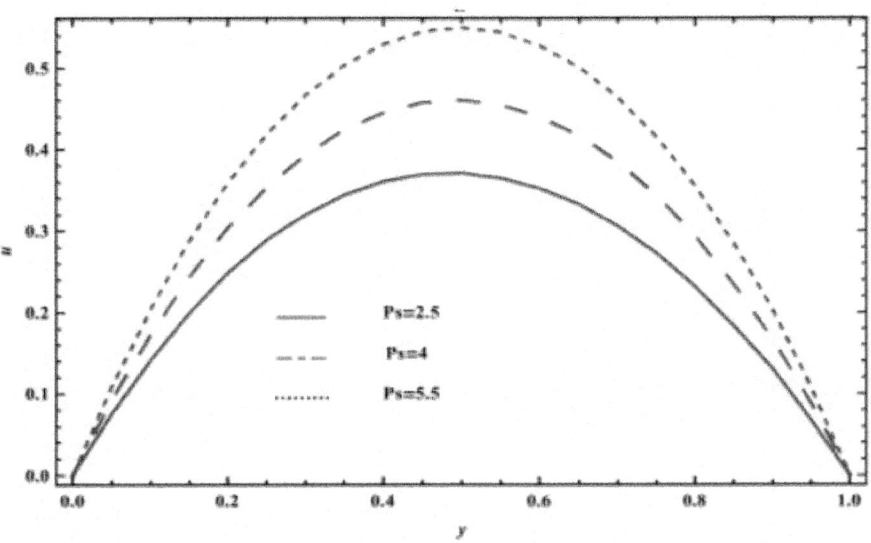

Figure 6: The velocity distribution is plotted versus y, for different values of Ps and for a system which has the particular values k_0=1.5, Re =1.6, Da=0.5, M=3, P_0=0.5, $\omega=\pi/4$, Pr=1, G_T=3.5, G_C=1.5, Sc=1.5, Sr=1, t=0.5

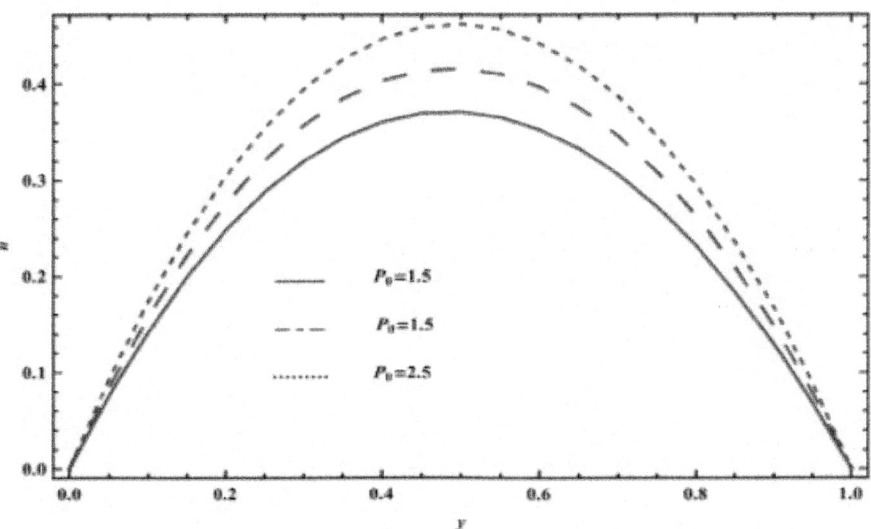

Figure 7: The velocity distribution is plotted versus y, for different values of P_0 and for a system which has the particular values k_0=1.5, Re =1.6, Da=0.5, M=3, Ps=2.5, $\omega=\pi/4$, Pr=1, G_T=3.5, G_C=1.5, Sc=1.5, Sr=1, t=0.5

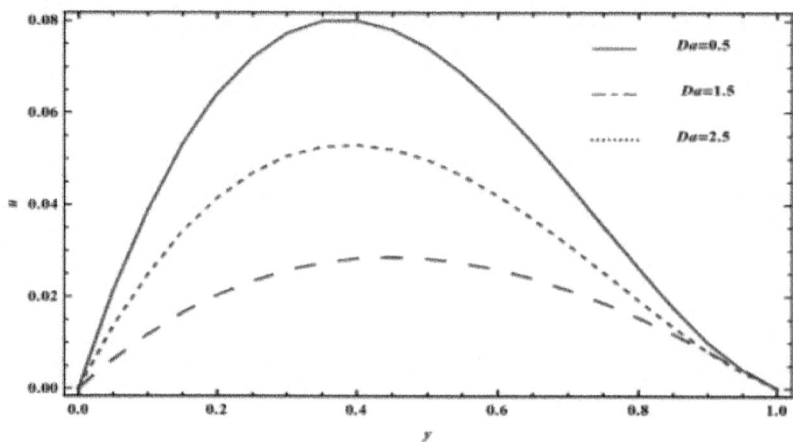

Figure 8: The velocity distribution is plotted versus y, for different values of Da and for a system which has the particular values k_0=1.5, Re =1.6, M=3, Ps=2.5, P_0=0.5, ω=π/4, Pr=1, G_T=3.5, G_C=1.5, Sc=1.5, Sr=1, t=0.5

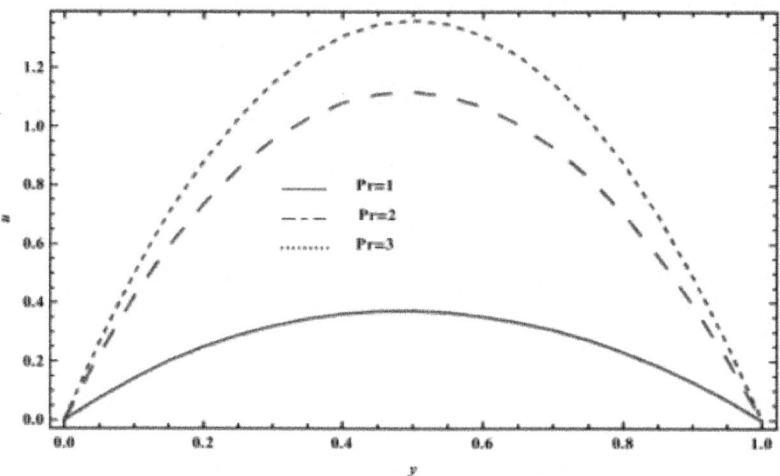

Figure 9: The velocity distribution is plotted versus y, for different values of Pr and for a system which has the particular values k_0=1.5, Re =1.6, Da=0.5, M=3, Ps=2.5, P_0=0.5, ω=π/4, G_T=3.5, G_C=1.5, Sc=1.5, Sr=1, t=0.5

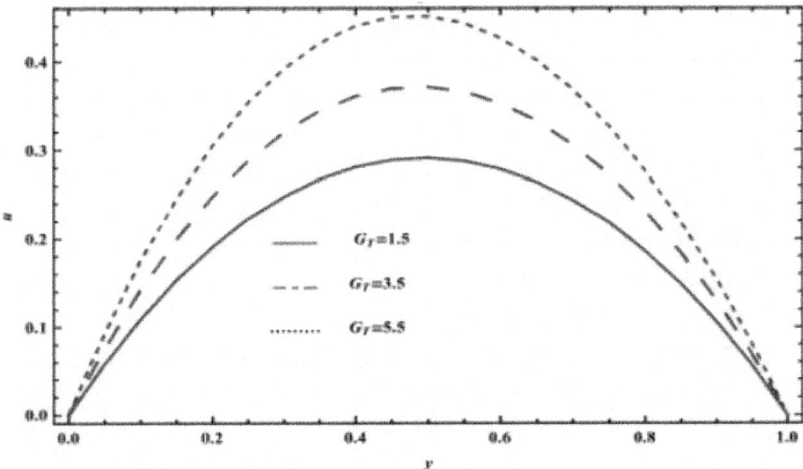

Figure 10: The velocity distribution is plotted versus y, for different values of G_T and for a system which has the particular values $k_0=1.5$, Re $=1.6$, Da=0.5, M=3, Ps=2.5, $P_0=0.5$, $\omega=\pi/4$, Pr=1, $G_C=1.5$, Sc=1.5, Sr=1, t=0.5

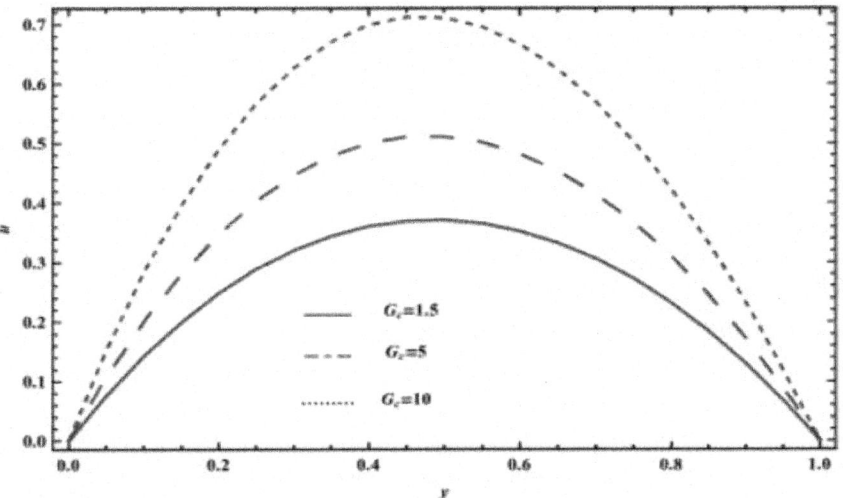

Figure 11: The velocity distribution is plotted versus y, for different values of G_C and for a system which has the particular values $k_0=1.5$, Re $=1.6$, Da=0.5, M=3, Ps=2.5, $P_0=0.5$, $\omega=\pi/4$, Pr=1, $G_T=3.5$, Sc=1.5, Sr=1, t=0.5

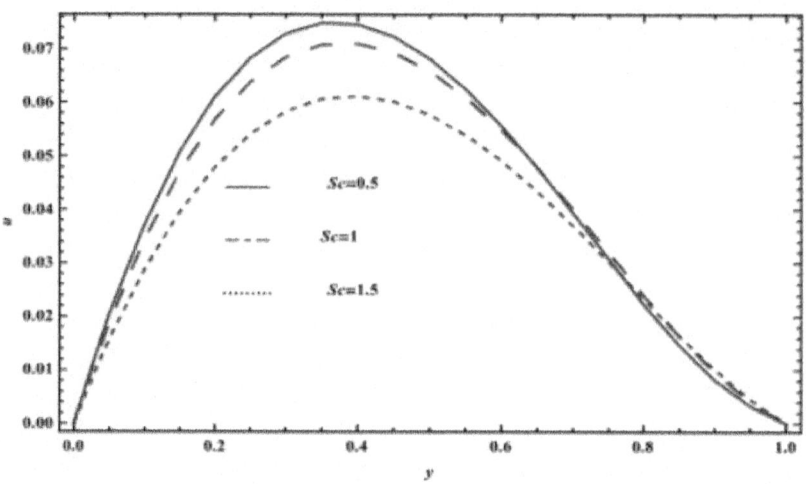

Figure 12: The velocity distribution is plotted versus y, for different values of Sc and for a system which has the particular values k_0=1.5, Re =1.6, Da=0.5, M=3, Ps=2.5, P_0=0.5, ω=π/4, Pr=1, G_T=3.5, G_C=1.5, Sr=1, t=0.5

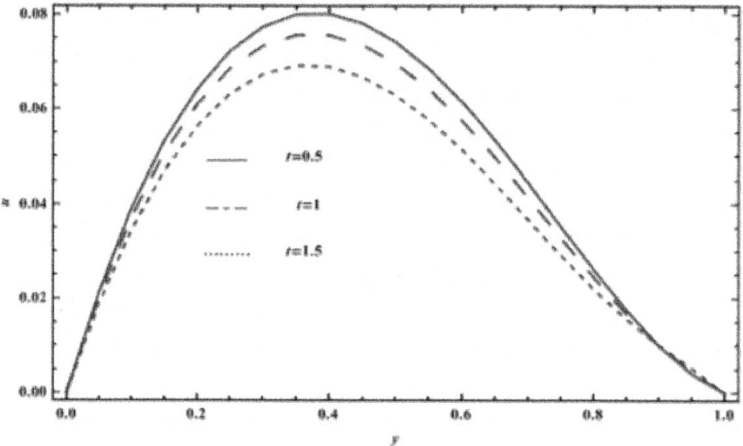

Figure 13: The velocity distribution is plotted versus y, for different values of t and for a system which has the particular values k_0=1.5, Re =1.6, Da=0.5, M=3, Ps=2.5, P_0=0.5, ω=π/4, Pr=1, G_T=3.5, G_C=1.5, Sc=1.5, Sr=1

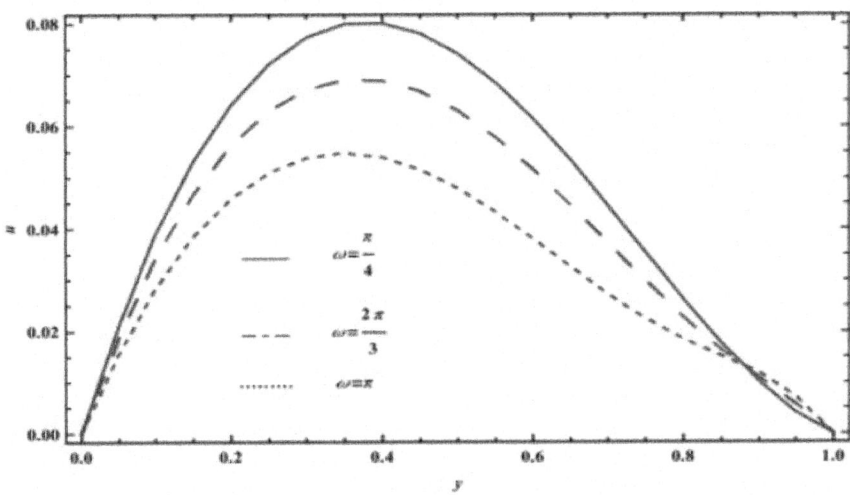

Figure 14: The velocity distribution is plotted versus y, for different values of ω and for a system which has the particular values $k_0=1.5$, Re =1.6, Da=0.5, M=3, Ps=2.5, $P_0=0.5$, Pr=1, $G_T=3.5$, $G_C=1.5$, Sc=1.5, Sr=1, t=0.5

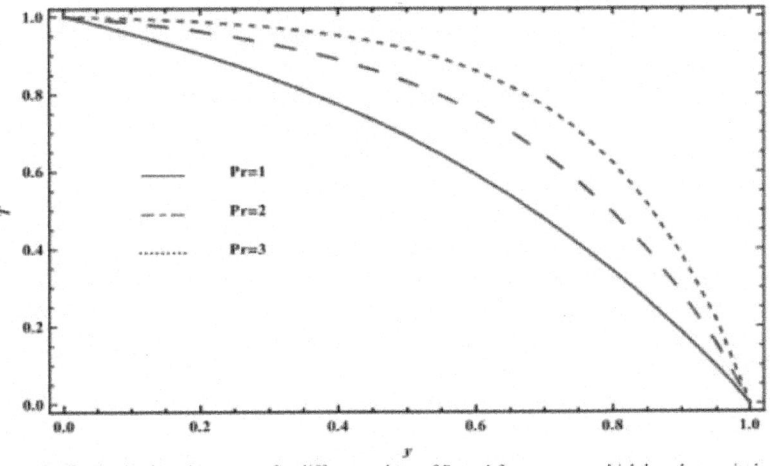

Figure 15: The temperature distribution is plotted versus y, for different values of Pr and for a system which has the particular values $k_0=1.5$, Re =1.6, Da=0.5, M=3, Ps=2.5, $P_0=0.5$, $\omega=\pi/4$, $G_T=3.5$, $G_C=1.5$, Sc=1.5, Sr=1, t=0.5

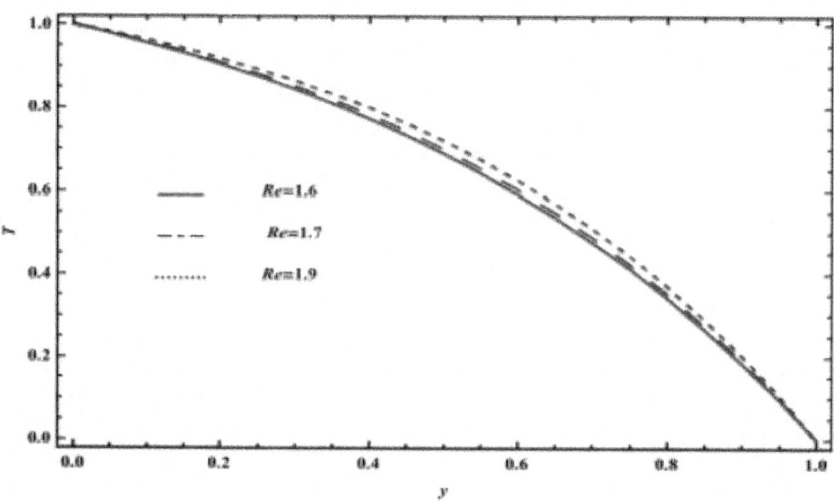

Figure 16: The temperature distribution is plotted versus y, for different values of Re and for a system which has the particular values $k_0=1.5$, Da=0.5, M=3, Ps=2.5, $P_0=0.5$, $\omega=\pi/4$, Pr=1, $G_T=3.5$, $G_C=1.5$, Sc=1.5, Sr=1, t=0.5

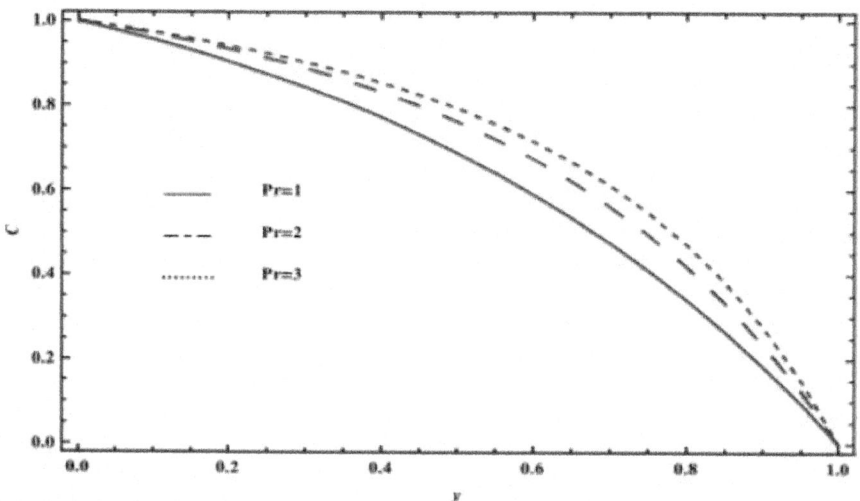

Figure 17: The concentration distribution is plotted versus y, for different values of Pr and for a system which has the particular values $k_0=1.5$, Re =1.6, Da=0.5, M=3, Ps=2.5, $P_0=0.5$, $\omega=\pi/4$, $G_T=3.5$, $G_C=1.5$, Sc=1.5, Sr=1, t=0.5

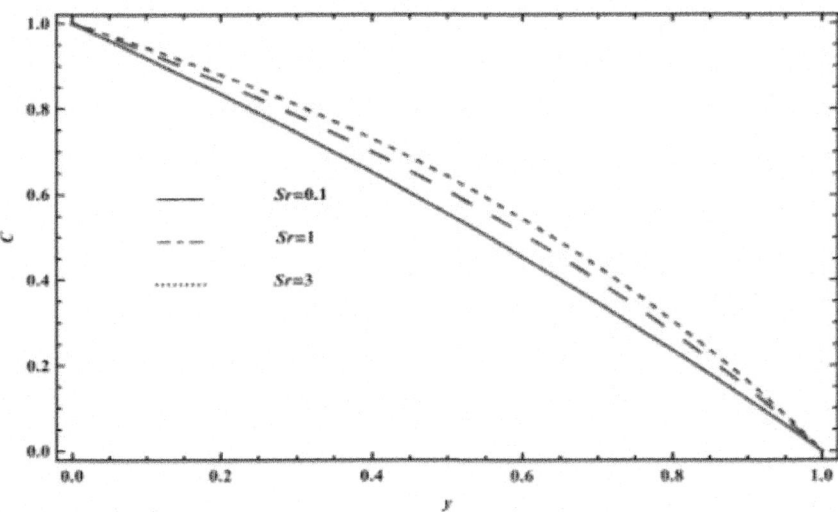

Figure 18: The concentration distribution is plotted versus y, for different values of Sr and for a system which has the particular values k_0=1.5, Re =1.6, Da=0.5, M=3, Ps=2.5, P_0=0.5, ω=π/4, Pr=1, G_T=3.5, G_C=1.5, Sc=1.5, t=0.5

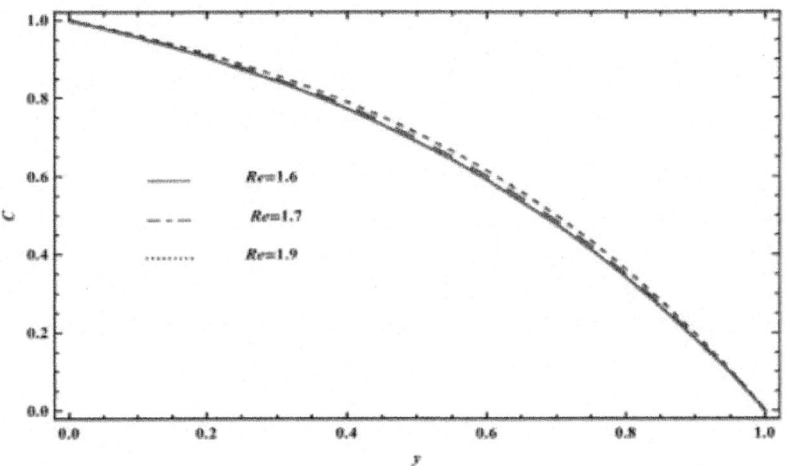

Figure 19: The concentration distribution is plotted versus y, for different values of Re and for a system which has the particular values k_0=1.5, Da=0.5, M=3, Ps=2.5, P_0=0.5, ω=π/4, Pr=1, G_T=3.5, G_C=1.5, Sc=1.5, Sr=1, t=0.5

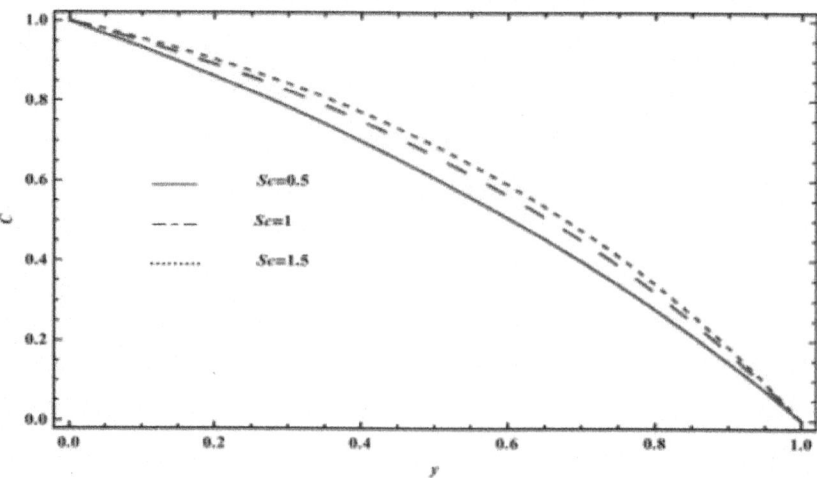

Figure 20: The concentration distribution is plotted versus y, for different values of Sc and for a system which has the particular values k_0=1.5, Re =1.6, Da=0.5, M=3, Ps=2.5, P_0=0.5, ω=π/4, Pr=1, G_T=3.5, G_C=1.5, Sr=1, t=0.5

The effects of the physical parameters on the velocity distribution are shown in figures 2 - 13. In these figures the velocity distribution u is plotted versus the coordinate y. Figures 2 and 3, illustrate the effects of the parameter of the non-Newtonian fluid k_0 and Reynolds number Re respectively. It is found that the velocity increases with increasing Re, but it decreases with increasing k0. We also noted that the velocity u increases with y till a definite value y = y_0 (represents the maximum u) and it decreases afterwards. This maximum value of u increases by increasing Re while it decreases by increasing k_0. Shawky [19] showed that the velocity increases with the increase of Reynolds number. The effect of the magnetic parameter M on the velocity is shown in figure 4, and it is shown that the velocity u increases by increasing M in the range of y shown in the figure, and also, the velocity increases with y, till a maximum value (at a finite value of y : y = y_0) after which it decreases. It is clear that the maximum of u increases by increasing M and this also occurs at another value y > y_0. The variation of u with y for different values of Soret number Sr is drawn in Fig. 5. It is observed that the behavior of the curves is the same as that obtained in Fig. 4. The obtained curves are very close to those obtained in Fig. 4, and they coincide when $0 \leq y \leq 0.1$, and afterwards. u increases by increasing Sr.

In figures 6 and 7, the graphs of the velocity of the fluid have been drawn against y for different values of Ps and P_0, respectively. It is noticed that the behavior of the curves are the same as that obtained in Fig. 3.The effects of

Darcy number Da and Prandtl number Pr on the velocity are elucidated in figures 8 and 9, respectively. It reveals that the velocity increases with increasing Pr, but it decreases with increasing Da. Figures 10 and 11, show the distribution of the velocity for various values of the temperature Grashof number G_T and concentration Grashof number G_C, Respectively. It is obvious that the velocity increases with the increase of both G_T and G_C. Also, the velocity increases with y , till a maximum value (at a finite value of y : $y = y_0$) after which it decreases.

The variations of the velocity distribution versus the coordinate y for various values of Schmidt number Sc is displayed in Fig. 12. The graphical results of Fig. 12 show that the velocity of fluid increases as Sc increases. While for $0.7 \leq y \leq 1$, near the right plate, the velocity decreases with the increase of Sc. The distributions of u within the coordinate y for various values of the time t and the frequency of the oscillating plate ω are exhibited in Figs. 13 and 14, respectively. From these figures, we observed that the effect of t and ω on u is opposite to the effect of Sc on u illustrated in Fig. 12, with the only difference that the obtained curves are very close to those obtained in Fig. 12.

The effects of Prandtl number Pr and Reynolds number Re on the temperature distribution T are indicated graphically in figures 15 and 16. In these figures, we observe that the temperature distribution increases with the increase of both Re and Pr. Also, the obtained curves in Fig. 16 are very close to those obtained in Fig. 15. The results in Fig. 15 and 16 are in agreement with those obtained by Shawky [19].

Figs. 17 and 18 show the behavior of the concentration C with the coordinate y for different values of Prandtl number Pr and Soret number Sr respectively. It is found that the effects of both Pr and Sr are to increase the concentration. Also, both Pr and Sr affect the relation between C and y. This relation is approximately linear at small values of Pr and Sr. In Figs. 19 and 20, the effects of Reynolds number Re and Schmidt number Sc on the concentration C respectively are presented. It is observed that the effect of Re and Sc on C is similar to the effect of Pr and Sr on C.

CONCLUSIONS

In this paper, we have studied the problem of MHD unsteady flow with heat and mass transfer of non-Newtonian fluid which is obeying the rheological equation of state due to Ree-Eyring's stress-strain relation. The flow is through a uniform porous medium between two vertical permeable parallel plates in the presence of convective heat and mass transfer. The equations of momentum, energy and concentration are solved analytically by using Lightill method [20]. The velocity, temperature and concentration distributions are obtained. The effects of various physical parameters of the problems on these distributions are discussed and illustrated graphically through a set of figures. Hence, this paper deals with an important branch of fluid mechanics, which has many important applications in many fields, such as biology, medicine and chemistry and also in the space science e.g.:

- The rheology of blood has received much study. Blood is rheologically complex on two counts: it is a suspension because erythrocytes with characteristic dimensions of several micrometers are present in excess of 40 % vol. and the suspension fluid itself exhibits non-Newtonian behavior because of the presence of high molecular-weight protein. The importance of rheological properties of other body fluids is now recognized. In particular, the rheological response of mucous in respiratory system of both infants and adults are an important factor for proper respiratory behavior. The lubricating action of synovial fluid in joints is likewise, strongly dependent on rheological properties [19].

- For engineering purposes, one is more interested in the values of the velocity and heat transfer than in the shape of the velocity and temperature profiles. The results of this problem are of great interest in petroleum applications such as rotating machinery, lubrication technology, viscometry, computer storage devices, food processing, biochemical operations, transport in polymers and understanding and predicting blood flow properties in large arteries. The flow of the Petroleum through the Porous ground represents a good example of the motion of our fluid especially in the motion of the fluid in the earth's core. Also, there are many applications of this motion in many fields such Astrophysical, Plasma MHD, Metallurgical processes and Geophysical applications [18].

REFERENCES

1. A. Yakhot, M. Arad, G. Ben-Dor, Numerical investigation of a laminar pulsating flow in a rectangular duct, Int. J. Numer. Meth. Fluids 29, (1999), 935-950

2. F. Fedele, D. Hitt, R. D. Prabhu, Revisiting the stability of pulsatile pipe flow, Eur. J. Mech., B, Fluids, 24 (2005), 237–254.

3. M. Zamir, The Physics of Pulsatile Flow, Springer-Verlag, New York, 2000.

4. D. F. Young, F. Y. Tsai, Flow characteristics in models of arterial stenosis –I. Steady Flow, J. Biomechanics, 6, (1973), 395-410.

5. D. F. Young, F. Y. Tsai, Flow characteristics in models of arterial stenosis –II. Unsteady Flow, J. Biomechanics, 6, (1973), 554-559.

6. M. Siouffi, R. Pelisser, D. Farahifar, R. Rieu, The effect of unsteadiness on the flow through stenoses and Bifurcation, J.Biomechanics, 17, (1984), 299-315.

7. M. S. Mandal, S. Mukhopadhyay, G. C. Layek, Pulsatile flow of shear-dependent fluid in a stenosed artery - Theoret. Appl. Mech., 39 (3), (2012). 209–231.

8. M. Nakamura, T. Sawada, Numerical study on flow of non-Newtonian fluid through axi-symmetric stenosis, J .Biomech. Eng., 110, (1988), 247-26

9. N. T. Eldabe, S. M. Elmohandis, Pulsatile magneto hydrodynamic viscoelastic flow through a channel bounded by two permeable parallel plates, J. Phys. Soc. Jpn., 64 (11), (1995), 4165.

10. N. T. Eldabe, S. M. Elmohandis, Effect of couple stresses on pulsatile poiseuille flow, Fluid Dynamic Research 15 (1995), 313.

11. P. Bitla, T. Kandala, V. Iyengar, Pulsating flow of an incompressible micro polar fluid between permeable beds, Nonlinear Analysis: Modeling and Control, 18 (4), (2013), 399–411.

12. M. A. Abdelnaby, N. T. M. Eldabe and M. Y. Abou-zeid, Numerical study of pulsatile MHD non-Newtonian fluid flow with heat and mass transfer through a porous medium between two permeable parallel plates, Ind. J. Mech. Cont. & Math Sci, 1, (2006), 1-15.

13. P. D. S. Verma, D. U. Singh, K. Singh, Pulsatile blood flow of a micro deformable fluid, Wear, 71, (1981), 333–346.

14. J. C. Misra, S. K. Ghosh, A mathematical model for the study of blood flow through a channel with permeable walls, Acta Mech., 122, (1997), 137–153.

15. K. Vajravelu, K. Ramesh, S. Sreenadh, P. U. Arunachalam, Pulsatile flow between permeable beds, Int. J. Non-Linear Mech., 38, (2003), 999–1005.

16. L. Kumar, S. Narayanan, Analysis of pulsatile flow and its role on particle removal from surfaces, Chem. Eng. Sci., 65, (2010), 5582–5587.

17. B. Mallik, S. Nanda, B. Das, D. Saha, D. S. Das, K. Paul, Pulsatile flow of casson fluid in mild stenosed artery with periodic body acceleration and slip condition, Sch. J. Eng. Tech., 1(1), (2013), 27-38.

18. M. Y. Abou-zeid, Numerical solutions for heat generation effect on MHD pulsatile non-Newtonian fluid flow with convective heat transfer in a non-Darcian porous medium between two rotating cylinders, Bull. Cal. Math. J. 101, (2009), 531-55

19. H. M. Shawky, Pulsatile flow with heat transfer of dusty magneto

hydrodynamic Ree-Eyring fluid through a channel, Heat Mass Transfer, 45(10), (2009), 1261-1269.

20. M. J. Lighthill, Introduction to Fourier analysis and generalized functions, Bull. Amer. Math. Soc., 65 (4), (1959), 248-249.

Chapter 12

ASSESSMENT OF NATURAL RADIOACTIVITY AND ASSOCIATED RADIATION HAZARDS IN TOPSOIL OF SAVAR INDUSTRIAL AREA, DHAKA, BANGLADESH

B. M. R. Faisal[1], M. A. Haydar[2], M. I. Ali[2], D. Paul[2], R. K. Majumder[3], M. J. Uddin[1]

[1]Department of Environmental Sciences, Jahangirnagar University, Dhaka-1342, Bangladesh

[2]Health Physics and Radioactive Waste Management Unit (HPRWMU), Institute of Nuclear Science and Technology (INST), Atomic Energy Research Establishment (AERE), Bangladesh Atomic Energy Commission (BAEC), Savar, Dhaka-1349, Bangladesh

[3]Nuclear Minerals Unit, Atomic Energy Research Establishment (AERE), Bangladesh Atomic Energy

ABSTRACT

The radioactivity levels of naturally occurring radionuclides ^{226}Ra, ^{232}Th and ^{40}K in thirteen topsoil samples, collected from Savar industrial area of Bangladesh, were measured by gamma ray spectrometry system using a High Purity Germanium (HPGe) detector of 40% relative efficiency. The calculated average activity concentrations of ^{226}Ra, ^{232}Th and ^{40}K in the collected samples were 23.31 Bq.kg^{-1}, 42.24 Bq.kg^{-1} and 733.19 Bq.kg^{-1}, respectively. The average activity concentration of ^{232}Th and ^{226}Ra in the present study is lower than that of the world-wide average value but the concentration for ^{40}K is much higher than the world average of 420 Bq.kg^{-1}. The estimated average absorbed dose rate (D), the outdoor annual effective dose (E) and the external hazard index (H$_{ex}$) were found as 67.08 nGyh^{-1}, 0.08 mSvy^{-1} and 0.38 Bq.Kg^{-1}, respectively. The absorbed dose rate and outdoor annual effective dose are slightly higher than the world average value but the external hazard index is lower than that of the recommended value. The results of present study show that the soil of the study area is safe from radiological hazards and will not pose any harmful effect to the environment. The outcome of this study may provide valuable information about radiation hazard as well as may be useful as the baseline data in the monitoring of environmental radioactivity in the industrial zone under study.

INTRODUCTION

Natural radioactivity is widespread in the earth environment and it exists in various geological formations such as earth crust, rocks, soils, plants, water and air [1]. In addition to natural sources, soil radioactivity is also affected by anthropogenic activities. Numerous types of human activities and non-nuclear industries contribute to further concentrate some of the natural radionuclides that can be found in the earth's crust affecting the human and the environment. The main contributors of radionuclides are ^{40}K and ^{232}Th and these radionuclides are not uniformly distributed in soils and vary from region to region [2]. Soil provides a direct source of radioactivity in food chain due to its uptake by agricultural plants. The radioactivity caused by radionuclides can transfer from soil, water and air to plants, trees and other biological elements and finally to human body which may deposited in this three media either by mining or any other natural or man-made activities. Gamma radiation emitted from primordial radionuclide and their progeny is one of the main external sources of radiation exposure to the humans [3].

Terrestrial radioactivity and the associated external exposure due to gamma radiation, depends primarily on the geological formation and soil type of the location; and these factors (geology and soil type) greatly influence the dose distribution from natural terrestrial radiation [1]. Majority of the external gamma dose rate above typical soils (95%) arises from primordial radionuclides incorporated in the soil [4]. The major potential hazard from the natural radiation is from external exposure either by direct exposure to soil or as they enter in many building materials.

Since natural radiation is the largest contributor of external dose to the world population, assessment of gamma radiation dose from natural sources is of particular importance. The radioactivity concentration of radionuclides above permissible level is very harmful for human health. Moreover, contamination may happen in the surrounding environmental elements such as soil due to the leaching of radionuclides from industrial activity. Industries cause environmental degradation thought the life cycle of a product starting from exploration of raw materials and energy resources to disposal of waste and end products [5]. The industrial units in Savar industrial area include garments, textile, knitting, leather goods, metal products, electronic goods, paper products, chemicals and fertilizers and miscellaneous products [6]. Industrial activities discharge untreated or poorly treated industrial wastewater, effluent and even sludge into the surrounding environment which may contain elevated level of radioactivity. Besides, the uneducated farmers in that area are randomly using fertilizers and pesticides in agricultural lands. Since, there is no data available on the radioactivity contents in the soil and radiological impact of the probable contents on people and environment in Savar industrial area. Therefore, the aim of the present study is to assess the probable natural terrestrial radionuclide contents and corresponding health risk due to exposure to these radionuclides present in the topsoil of Savar industrial area.

MATERIALS AND METHODS

Study Area

The study area lies between 2354'47.8" to 2348'33.1" north latitude and 9014'52.6" to 9014'44.5"east longitude (**Figure 1**). The industries along both sides of Dhaka-Aricha highway from Hemayetpur to Savar bus stand are in industrial cluster-6, 'Dhaka Export Processsing Zone Depz' (DEPZ) and DEPZ (extended), Ashulia and Jirabo industrial area are in cluster-8. The study area comprises of many isolated water bodies occupying the low lying and depressed areas connected to or out-of-the-way from the river system. The Bansi-Daleshwari and Turag river system comprise the drainage network of the study area–where the Bansi flows on the west and the Turag is away on the east. The land of Savar industrial area composed of Pleistocene red clay and recent alluvium soil. The major part of the land is used for the cultivation of agricultural products and the rest is used for industrial activity.

Sampling Locations

Soil samples were collected from different locations of the study area adjacent to Hemayetpur, Savar, Ashulia, Bipyle, Mirjanagar and Nayerhat by traverse method. This area receives the drainage residue of the industrial area and other adjoining residential and numerous classified/unclassified industrial wastes. The geographical location of each sampling points were determined with a handheld global positioning system (GARMIN). A total number of thirteen topsoil samples were collected in May, 2013 (**Figure 1**).

Processing of Soil Samples

After collection, packaging and marking all the samples were transported to and preserved at the sample processing laboratory of the Health Physics and Radioactive Waste Management Unit, Bangladesh Atomic Energy Commission (BAEC), Savar, Dhaka. Each sample was then cleaned and dried in the sun separately and crushed into fine powder by using a grinder. The samples were then mixed homogeneously and passed through a sieve of 200 μm mesh size. The homogenized soil samples were then dried in a temperature controlled oven at 110°C for about 24 hours. The samples were then transferred to sealable cylindrical plastic containers of 7 cm and 5.5 cm in diameter and the weights of the samples were recorded using an electrical balance. The sample-filled containers were marked individually with identification parameters e.g., sample ID, date of preparation and net weight. The containers were sealed tightly with insulating tape around their opening for impeding the possibility of moisture contamination. In order to maintain radioactive secular equilibrium between ^{226}Ra and its daughter products, the sealed containers were stored for a period of 4 weeks [7].

Experimental Set-Up

The detection and measurement of radionuclides in the samples were carried out by gamma ray spectrometry system using a vertical co-axial cylindrical High Purity Germanium (HPGe) detector of 172 cm^3 active volume and with 40% relative efficiency. The p-type HPGe detector supplied by CANBERRA (Model- GC 4020) had a resolution of 2 keV at 1332 keV of Cobalt-60 gamma-ray line. The detector was coupled to a 16 k-channel analyzer. The spectra of all samples were perfectly analyzed using Genie-2000 spectra analysis software (which matched various gamma energy peaks to a library of all possible radionuclides) to calculate the concentrations of ^{226}Ra, ^{232}Th and ^{40}K. The detector was enclosed in a cylindrical shielding container made of Lead and Iron with 11.3 cm thickness, 51 cm height and 28 cm internal diameter and having a fixed bottom and moving cover to reduce the external gamma-ray background. All the samples were counted for 10 ks. Prior to the measurement of the samples, the environmental gamma background at the laboratory site was determined with identical plastic container used in the sample measurement. The energy regions selected for the corresponding radionuclides were 295 keV and 352 keV of ^{214}Pb and 609 keV, 1120 keV and 1764 keV for^{214}Bi for ^{226}Ra, 583 keV and 2614 keV of ^{208}Tl, 911 keV and 969 keV of ^{228}Ac for ^{232}Th and 1460 keV for ^{40}K [8].

Calibration of the Detector

The efficiency calibration of the detector was performed by standard sources of solid matrices prepared using^{226}Ra standard using identical containers used for the measurement of the samples, e.g., 180 ml plastic container for solid samples. The preparation process of standard sources had been reported elsewhere [9]. The detector efficiency calibration curves as a function of energy for solid matrices are shown in Figure 2. The energy calibration of the detector was performed by ^{137}Cs and ^{60}Co point sources.

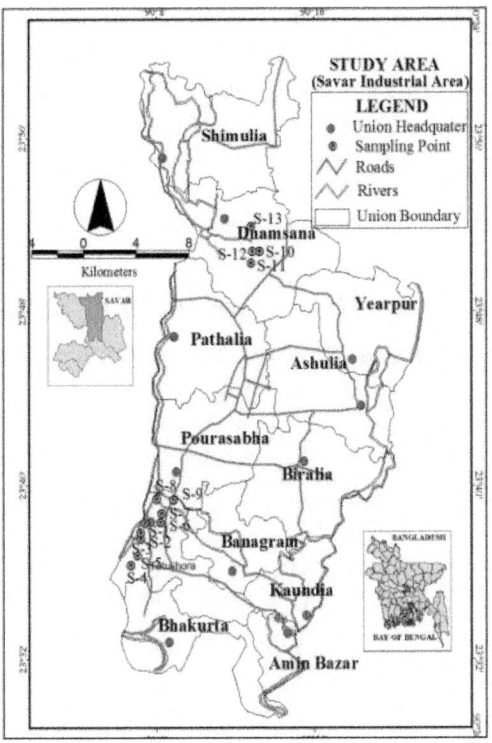

Figure 1: Location map of Savar industrial area, Bangladesh

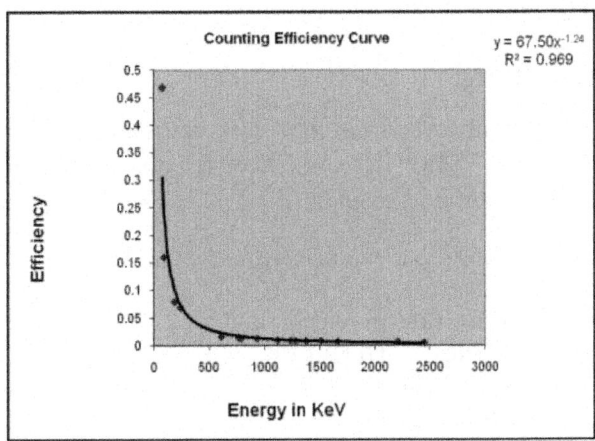

Figure 2: Efficiency curve of 40% relative efficiency for the solid matrix

Activity Concentrations of Soil

The radionuclide contents and their activity levels in the samples were measured using a calibrated HPGe detector. The activity concentration (A) of each radionuclide in the samples was determined by using the net count (cps) (found by subtracting the background counts from the gross counts with same counting time under the selected photo peaks), weight of the sample, the photo-peak efficiency and the gamma intensity at a specific energy as [10].

$$A = \frac{CPS}{E \times I \times W}$$
(1)

Where, A = Activity concentrations of the sample in Bq.kg^{-1},

CPS = the net counts per second = cps for the sample- cps for the background value,

E = the counting efficiency of the gamma energy,

I = Absolute intensity of the gamma ray and

W = Net weight of the sample (in kilogram).

The errors in the measurement were expressed in terms of standard deviation ($\pm 2\sigma$), where σ is expressed as [1].

$$\sigma = \left[\frac{N_s}{T_s^2} + \frac{N_b}{T_b^2} \right]^{\frac{1}{2}}$$
(2)

Where, N_s is the sample counts measured in time T_s, and N_b is the background counts measured in time T_b. The standard deviation $\pm 2\sigma$ in cps was converted into activity in Bq.kg^{-1} according to equation (1).

Absorbed Dose Rates

The external outdoor absorbed gamma dose rates due to terrestrial gamma rays from the nuclides ^{226}Ra,^{232}Th and ^{40}K at 1m above the ground level were calculated as follows outlined by [11].

$$D(nGy^{-1}) = 0.462C_{Ra} + 0.60C_{Th} + 0.042C_K$$
(3)

Where, D is the dose rate in nGyh^{-1} and C_{Ra}, C_{Th} and C_K are the specific activities (Bq.Kg^{-1}) of ^{226}Ra, ^{232}Th and ^{40}K, respectively.

Outdoor Annual Effective Dose

The absorbed dose rate was converted into annual effective dose equivalent by using a conversion factor of 0.7 SvGy^{-1} [1] and 0.2 for the outdoor occupancy factor by considering that the people on the average spent 20% of their time in outdoors [12]. The effective dose due to natural activity in the soil samples was calculated by [11].

$$\text{Effective dose rate}(mSvy^{-1}) = D(nGy^{-1}) \times 8760(hy^{-1})$$
$$\times 0.2 \times 0.7(SvGy^{-1}) \times 10^{-6} \tag{4}$$

Where, D = the absorbed dose rate in air (nGyh^{-1}); 8760 = the time in hours for one year; 0.2 = the outdoor occupancy factor; 0.7 = SvGy^{-1} is the quotient of effective dose equivalent rate to absorbed dose rate in air and 10^{-6} = the factor converting nano into milli.

External Hazard Index (H$_{ex}$)

Local soil of the area is used for the construction of houses and also for agricultural purposes. These soils may contribute to the external gamma dose rates to the public. The external hazard index (H$_{ex}$) is the indoor radiation dose rate due to the external exposure to gamma radiation in construction materials of dwelling which is calculated by [13].

$$H_{ex} = \frac{A_{Ra}}{370} + \frac{A_{Th}}{259} + \frac{A_K}{4810} \tag{5}$$

Where, A$_{Ra}$, A$_{Th}$ and A$_K$ are the activity concentration in Bq.Kg^{-1} of ^{226}Ra, ^{232}Th and ^{40}K respectively. The value of H$_{ex}$ must be lower than unity in order to keep the radiation hazard insignificant.

RESULTS AND DISCUSSIONS

Activity Concentration in Topsoil Samples

The results of activity concentrations of radionuclides obtained from gamma spectrometry technique for 13 topsoil samples collected from Savar industrial area are presented in **Table 1**, with the uncertainty level of ±2σ. The results for the radionuclides ^{226}Ra, ^{232}Th and ^{40}K are also shown graphically in **Figure 2**, **Figure 3** and **Figure 4**, respectively.

Table 1: Activity concentration of ^{226}Ra, ^{232}Th and ^{40}K in topsoil sample

Sample ID	Activity concentration (Bq.Kg^{-1})		
	^{226}Ra	^{232}Th	^{40}K
S-01	31.28 ± 6.50	46.44 ± 13.13	848.67 ± 171.60
S-02	6.31 ± 7.79	47.78 ± 13.03	894.64 ± 170.92
S-03	18.63 ± 6.35	41.11 ± 12.98	945.21 ± 171.66
S-04	14.47 ± 6.66	32.76 ± 10.51	439.84 ± 169.67
S-05	14.98 ± 6.39	20.38 ± 12.86	591.24 ± 171.35
S-06	32.05 ± 6.42	46.20 ± 12.94	1064.82 ± 171.17
S-07	34.56 ± 7.02	41.17 ± 10.61	798.89 ± 172.16
S-08	14.98 ± 6.28	23.37 ± 12.70	666.81 ± 169.10
S-09	34.78 ± 6.57	54.54 ± 10.73	826.57 ± 171.91
S-10	7.58 ± 6.27	27.02 ± 12.91	556.53 ± 170.48
S-11	21.30 ± 8.14	32.90 ± 8.14	666.81 ± 174.06
S-12	47.78 ± 7.07	67.57 ± 10.75	723.49 ± 170.48
S-13	24.34 ± 6.57	67.95 ± 11.18	507.96 ± 171.23
Maximum	47.78 ± 8.14	67.95 ± 13.13	1064.82 ± 174.06
Minimum	6.31 ± 6.27	20.38 ± 8.14	439.84 ± 169.10
Average	**23.31 ± 6.77**	**42.24 ± 11.73**	**733.19 ± 171.22**
World Average	33	45	420

Activity Concentration of ^{226}Ra

The concentration of ^{226}Ra ranges from 6.31 ± 6.27 Bq.Kg^{-1} to 47.78 ± 8.14 Bq.Kg^{-1} with an average of 23.31 ± 6.77 Bq.Kg^{-1}. The highest value for ^{226}Ra (47.78 ± 8.14) was found in S-12, near the Berger paints industry, Ashulia and lowest value (6.31 ± 6.27) was found in S-09, beside the Dhelasware river bank, Hemayetpur(**Figure 2**). The average radioactivity level for ^{226}Ra (23.31 Bq.Kg^{-1}) is less than the worldwide average value of 33 Bq.Kg^{-1} [1].

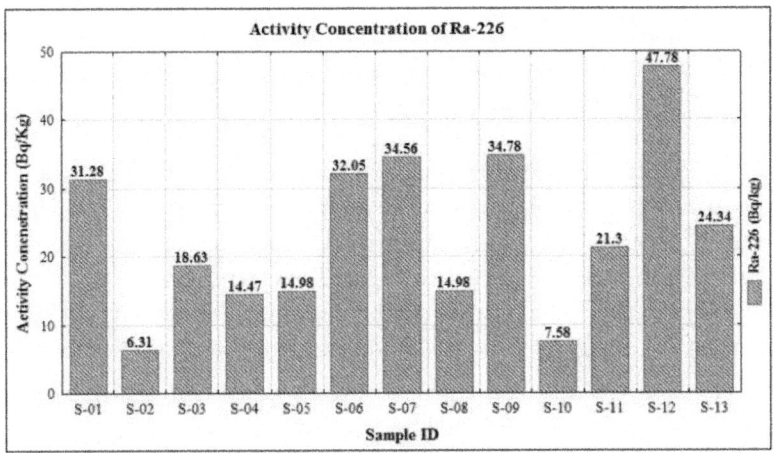

Figure 2: Distribution of [226]Ra in topsoil samples

Activity Concentration of [232]Th

The [232]Th radioactivity concentration varies from 20.38 ± 8.14 to 67.95 ± 13.13 with an average of 42.24 ± 11.73 Bq.kg[-1]. The highest [232]Th activity of 67.95 ± 13.13 Bq.kg[-1] was found in the S-13 near the Berger paints industry, Ashulia while the lowest [232]Th activity concentration of 20.38 ± 8.14 Bq.kg[-1] was found in S-05 Karnapara, Savar **(Figure 3)**. The average radioactivity level of [232]Th (42.24 Bq.kg[-1]) is also lower than the worldwide average value of 45 Bq.kg[-1] [1].

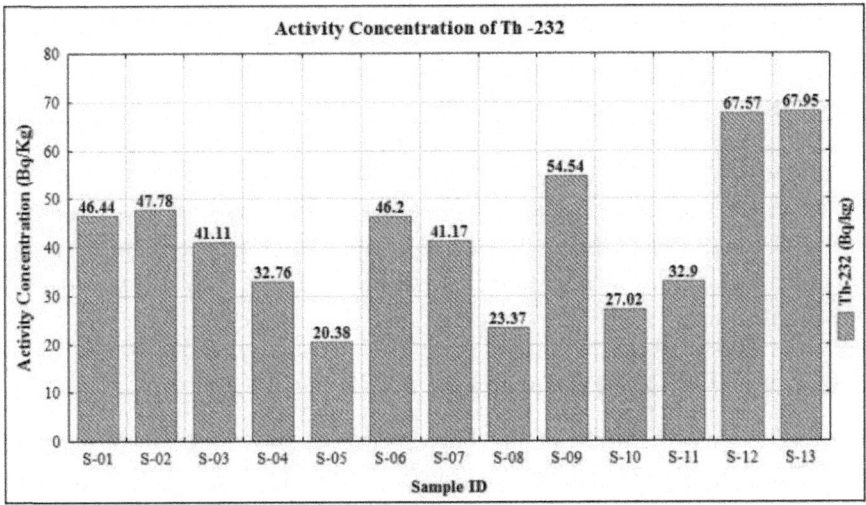

Figure 3: Distribution of [232]Th in topsoil samples

Activity Concentration of ^{40}K

The activity concentrations of ^{40}K ranges from 439.84 ± 169.10 to 1064.82 ± 174.06 Bqkg^{-1} with an average value of 733.19 ± 171.22 Bqkg^{-1}. The highest ^{40}K concentration of 1064.82 ± 174.06 Bqkg^{-1} was found in S-04 beside the Anlima dying industry and the lowest value of 439.84 ± 169.10 Bqkg^{-1} was found in S-04 karnapara, Savar **(Figure 4)**. The average value of ^{40}K 733.19 Bqkg^{-1} is significantly higher than that of the worldwide average value of 420 Bqkg^{-1} [1]. This result strongly suggests that radioactivity level for ^{40}K may arise from Madhupur clay formation as well as agricultural activity which requires potassium enrich fertilizers and pesticides and various industrial inputs in this area.

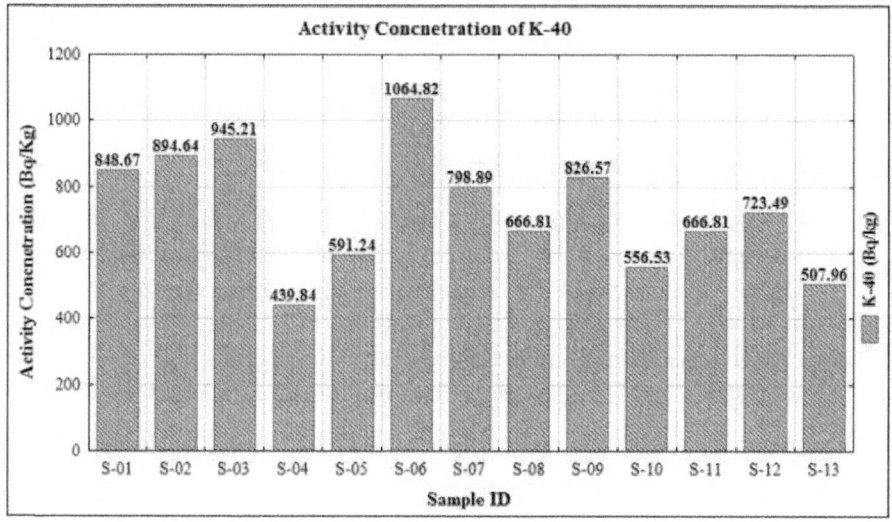

Figure 4: Distribution of ^{40}K in topsoil samples

The results of the current study reveal that the activity concentration of ^{226}Ra is lower than the world average level whereas the activity concentration of ^{232}Th is much closer the world average value. On the other hand, the activity concentration of ^{40}K in soil samples is much higher than the world average concentration.

Radiological Hazard Assessment

Different known radiation health hazard indices have been used in radiation studies to arrive at a better and safer conclusion on the health status of an exposed person and environment now a day. In order to assess the health effects, the 'Absorbed Dose Rate' (D), 'Outdoor Annual Effective Dose' (E) and 'External

Hazard Index' (H_{ex}) have been calculated from the activity concentrations of ^{226}Ra, ^{232}Th and ^{40}K using equations (3), (4) and (5), respectively and the values are shown in **Table 2**.

Table 2: Absorbed Dose Rate (D), Outdoor Annual Effective Dose (E) and External Hazard Index (H_{ex}) of all topsoil samples

Sample ID	Absorbed Dose Rate, D (nGyh⁻¹)	Outdoors Annual Effective Dose, E (mSvyr⁻¹)	Hex (Bq.Kg⁻¹)
S-01	78.15	0.10	0.44
S-02	69.35	0.09	0.39
S-03	73.14	0.09	0.41
S-04	44.94	0.06	0.26
S-05	44.06	0.05	0.24
S-06	87.43	0.11	0.49
S-07	74.38	0.09	0.42
S-08	49.04	0.06	0.27
S-09	83.72	0.10	0.48
S-10	43.20	0.05	0.24
S-11	57.72	0.07	0.32
S-12	93.27	0.11	0.54
S-13	73.62	0.09	0.43
Maximum	93.27	0.11	0.54
Minimum	43.20	0.05	0.24
Average	**67.08**	**0.08**	**0.38**
World Average	59	0.07	1.00

The results shown in Table 2 depict that the absorbed dose rates due to the terrestrial gamma rays at 1m above from the ground are in the range of 43.20 to 93.27 nGyh⁻¹ with an average of 67.08 nGyh⁻¹ for topsoil samples in the area under study. This value is slightly higher than the world average value of 59 nGyh⁻¹ [1].

The outdoor annual effective doses are in the range of 0.05 to 0.11 mSvyr⁻¹ with an average of 0.08 mSvyr⁻¹ in topsoil samples; which is comparable with the world average value of 0.07 mSvyr⁻¹ [1].

On the other hand, the external radiation hazard index ranges from 0.24 to 0.54 Bq.Kg⁻¹ with an average value of 0.38 Bq.Kg⁻¹, which is far less than the unity indicating the non-hazardous category of the samples. Therefore, it can be concluded that the radiation hazard due to the radiation exposure to the soils under investigation is insignificant. The values of hazard indices confirm that the study area is safe to carry out industrial and agricultural activities for the workers and general public and moreover, no significant radiological impacts have been observed on the surrounding environment.

CONCLUSIONS

The radionuclide contents, activity concentrations and radiological impact of the topsoil samples collected from the Savar industrial area of Bangladesh were investigated in the present study. The natural radioactivity concentrations of ^{226}Ra and ^{232}Th were slightly lower than that of the world average values for topsoil samples. But the radioactivity concentration of ^{40}K was significantly higher than the world average value. However, slight variation in the radioactivity content in soil observed with different locations mainly due to soil type, formation and transport process involved in the study area. The values of average absorbed dose rate and outdoor annual effective dose were slightly higher than the world average values; whereas the external hazard indices were found less than unity which indicates that there is no probability of immediate health effect on workers and public due to natural radioactivity present in the samples of the study area. Therefore, the area under present study may be termed as radiolologically safe. Nevertheless, elevated level of health risk may be caused due to natural terrestrial radiation from these soils on the inhabitants of this area if the uncontrolled industrial process and agricultural activities continue.

ACKNOWLEDGEMENTS

The author would like to acknowledge the Ministry of Science and Technology, The Peoples' Republic of Bangladesh for the NST fellowship to carry out the research. The authors also like to thank the technical support staffs of Health Physics and Radioactive Waste Management Unit (HPRWMU), INST, Atomic Energy Research Establishment, Savar, Dhaka and Department of Environmental Sciences, Jahangirnagar University, Savar, Dhaka-1342 for their help in conducting the study.

REFERENCES

1. UNSCEAR, (United Nations Scientific Committee on the Effects of Atomic Radiation Sources) Effects and risks of ionizing radiation, (2000).

2. Miah, F. K., Roy, S., Touhiduzzaman, M. & Alam, B. Distribution of radionuclides in soil samples in and around Dhaka city. Applied Radiation and Isotopes, 49 (1, 2), pp. 133-137 (1998).

3. UNSCEAR, (United Nations Scientific Committee on the Effects of Atomic Radiation) Sources, Effects, and Risks of Ionizing Radiation, (1993).

4. Jabbar, A., Tufail, M., Arshed, W., Bhatti, A. S., Ahmad,S. S., Akhter, P. and Dilband, M. Transfer of radioactivity from soil to vegetation in Rechna Doab, Pakistan. Isotopes in Environmental and Health Studies, 46, pp. 495 (2010).

5. Suzuki, K.T.H., Sunaga, A.Y., Hatakeyama, Y., Sumi and Suzuki. Binding of cadmium and copper in the mayfly baetis thermicus larvae. Comp. Biochem. Physical, 91c:487-492 (1998).

6. Khan, M.K., Alam, A.M., Islam, M.S., Hassan, M.Q., Al Mansur, M.A. Environmental pollution around Dhaka EPZ and its impact on surface and groundwater, Bangladesh. Journal Sci. Ind. Res. 46, pp. 153–162 (2011).

7. Hasan, M. M., Ali, M. I., Paul, D., Haydar, M. A. and Islam, S. M. A. Measurement of Natural Radioactivity in Coal, Soil and Water Samples Collected from Barapukuria Coal Mine in Dinajpur District of Bangladesh. Journal of Nuclear and Particle Physics, 3(4): 63-71 (2013).

8. Roessier, C. E., Smith, Z. A., Bloch, W.E. and Prince, R.J. Uranium and radium in Floride phosphate materials, Health Physics. vol. 37, pp. 269-277 (1970).

9. Usif, M. A. and Taher, A. E. Radiological assessment of Abu-Tartur phosphate, western desert Egypt, Radiation Protection Dosimetry, vol.130, pp. 228-235 (2008).

10. Knoll, G. F. Radiation detection and measurement. 2nd edition, John Wiley and Sons, Inc.pp.388-89 (1989).

11. Kessaratikoon, P. and Awaekechi, S. Natural radioactivity measurement in soil samples collected from municipal area of Hat Yai District in Songkhla Province, King Mongkut's Institute of Technology Ladkrabang Science Journal, Vol. 8, No. 2, , pp. 52-58 (2008).

12. Debertin, K., and Helmer, R.G. Gamma and X-ray spectrometry with semiconductor detectors. Elsevier Science, Amsterdam (1988).

13. Lu, X. and Xiolan, Z. Measurement of natural radioactivity in sand.

Chapter 13

DEVELOPMENT OF THE STABILITY ON THE LASER SYSTEM USED AT SATELLITE LASER RANGING STATION

H. M. Mansour[1], A. A. I. Khalil[2], M. Y. Helali[3], M. Mansour[2]

[1]Department of Physics, Faculty of Science, Cairo University, Giza, 12613, Egypt

[2]National Institute of Laser Enhanced Sciences, (NILES), Cairo University, Giza, 12613, Egypt

[3]National Research Institute of Astronomy and Geophysics, Helwan, Cairo, Egypt

ABSTRACT

A level of the Nd:YAG oscillator stability has been experimentally determined for different setups of resonator Q-switches. Performed experiments quantify an influence of an active modulation element on generated laser pulses stability. Different types of saturable absorbers were investigated and improved stability of double switching (active mode locker plus passive saturable absorber mode-locker) was confirmed. The best stability has been achieved with the aid of dual switching of an acousto-opticmodulator and a saturable absorber-dye solution 3955 in ethylalcohol to be used for satellite laser radar station at Helwan city in Egypt. The energy stability of the laser output can improve the accuracy and reliability of the satellite laser ranging station. The oscillator output picosecond pulses resulting from both the saturable absorber dye and three different multiple quantum well (semiconductor) saturable absorbers have been compared.Experimental results revealed that thedye ML51 mode-locking was suitable for Q-switched and mode-locked solid-state lasers.

INTRODUCTION

High-power, pulsed Nd: YAG laser is an attractive and promising system for many applications such as switching high voltage electric discharges, micromachining, micro- structuring, materials processing technologies, electronics and spectroscopy[1,2]. Active mode-locking involves the periodic modulation of the cavity losses or of the round-trip phase change, achieved e.g. with an acousto-optic or electro-optic modulator, a semiconductor electro-absorption modulator. If the modulation is synchronized with the cavity round trips, this leads to thegeneration of ultra-short pulses. The achieved pulse duration is governed by a balance of pulse shortening through the modulator[3,4] and pulse broadening via other effects, such as the limited gain bandwidth. Passive mode-locking (with a saturable absorber inside the laser resonator), it allows generating much shorter (femtosecond) pulses[5-7]. The pulse duration can be even well below the recovery time of the absorber; however passive mode-locking techniques for the generation of ultra-short pulse trains are preferred over active techniques due to the ease of incorporation of passive devices into various laser cavities. The liquid saturable absorber used for mode-locking and Q-switching is still considered to be the main drawback of passively mode-locked flash lamp pumped neodymium doped solid-state lasers[8-13]. In the following experimentthe output pulse trains are measured and compared between the use of only passive mode-locker, and the case where both passive (saturable dyes) and active mode-lockers (acousto-optic technique) are combined together. Also, the comparison between the mode-locked trains from multiple-quantum-well (MQW) No.758 and from liquid dye ML51 has been obtained.

EXPERIMENTAL SETUP

The oscillator of the described system works in mode-locking regime as shown in Fig. 1. In this regime particular longitudinal modes of electromagnetic field in the oscillator are synchronized and thus it's achieved ideally one shorter pulse in the oscillator. In this experiment common modulation of Q-factor was chosen, i.e. the modulation of losses in the resonator. To this reason, the special saturable absorber with short relaxation time can be used which, from certain intensity of absorbed energy, suddenly increases its transmission and thus decreases losses in the resonator. There exists also the second possibility of modulation, the acousto-optic modulator, a crystal, in which a periodical structure is established with the aid of applied high frequency field. On this structure the laser generated radiation is deflected.

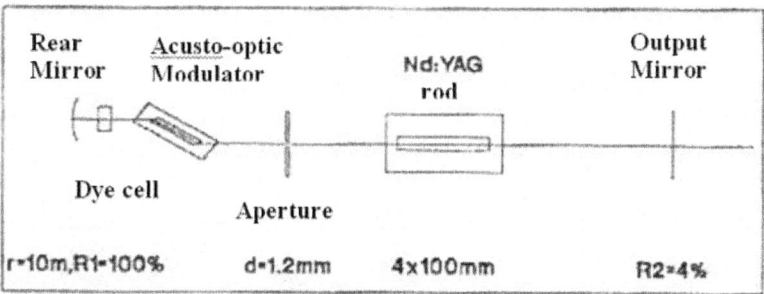

Figure 1: Setup for the passive-active Q-switching of Nd: YAG laser oscillator

By control of intensity of this field, a regulation of the Q-factor of the resonator can be reached. The open resonator of the oscillator consists of two mirrors: a high reflectance R_1=100%, dielectric concave mirror M_1 has a radius of curvature of r=10 m and a wedged mirror M_2, which has a reflectance of 4% and its outer side has an antireflective coating for generated wavelength. An Nd:YAG crystal (4 x 100 mm) and a xenon flash lamp (7 x 75 mm) were placed in single-elliptical cavity with silver inner coating. A 1.2 mm shutter ensures a generation in a basic transverse mode. The resonator Q-factor is switched by the saturable absorber or double by the saturable absorber and the acousto-optic modulator. The resonator length was adjusted for acousto-optic modulator frequency of 94.75 MHz. The saturable absorber is made up of dye solution with such initial transmission which ensures generation of one burst of pulses. These solutions were used: ML51 in dichloroethan, 3955 in ethylalcohol and ML63 in ethylalcohol. This passive absorber was placed in a cell 5 mm thick near the high reflection mirror. The acousto-optic modulator was located in front of the cuvette. Its faces are cut in Brewster angle. This brings additional losses for one polarization (vertical to the scheme plane) and generated radiation is linearly polarized. In experiment with the aid of piezoelectric joule-meter (Gentec PRJ-M) measured output laser energy and with an oscilloscope (Tektronix TDS 350) voltage pulses are observed. Amplitude of these pulses is proportional to energy, which is in the burst of pulses. Response time of this detector is 25 ms, which ensures capturing of whole burst energy. With repetition rate of 5 Hz it doesn't come to bursts interference. Nd:YAG laser system in Helwan SLR station consists of five parts: oscillator, pulse selector, amplifier chain (three Nd: YAG amplifiers), second harmonic generator, and the output telescope. Each of which has its own explanation in the following subsections. Figure 2 shows the different parts of the whole laser system. The flash lamp oscillator circuit is built from 120 µF capacitor bank and 22 µH inductance. In a free running mode, the cavity lasing threshold is 40 Joules; the mode-locking operating threshold is 45 joules. The pinhole diameter (1.6 mm) is placed near the rear concave mirror, and is working as a loss against the oscillation of transverse modes inside the cavity except of the TEM_{00} mode. The dye cell contains organic dye, e.g., ML51 dissolved in dichlorethan and it is used for Q-switching and mode-locking, the dye cell is placed next to the rear mirror because in this place it has the largest exposure volume to the laser radiation and this makes it to have a long life and not to be dissociated rapidly due to intense laser radiation, the dye cell contains 2 cc of dye solution. After the laser pulses exit from the oscillator output mirror they are reflected on the mirror M_1, which is installed in the vertical plane at 45° with respect to the resonator axis, and then on M_2 to pass through a telescope T_1 to enlarge the spot diameter 3 times and a half wave plate which changes the horizontal polarization direction of the oscillator output 90 degrees, then we obtain vertically polarized mode-locked train of pulses which reflect on M_3. The pulse selector consists of two Pockels cells installed between two crossed polarizer's.

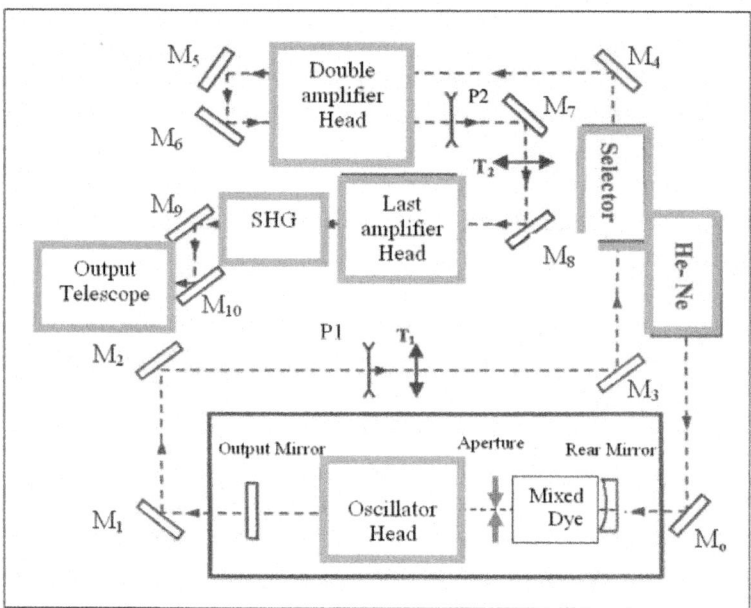

Figure 2: Block diagram of the mode-locked Nd:YAG laser system: M_o - silver total reflecting mirror for the He-Ne laser, M_1-M_8 - total dielectric reflecting mirror for 1.06 μm radiation, M_9 - dichroic mirror which reflects 0.53 μm and transmits 1.06 μm radiations, M_{10} - dielectric mirror totally reflecting radiation with the wavelength 0.53 μm

Normally, no voltage is applied on these two cells, and therefore the train of pulses of vertical polarization is reflected from its direction by the 2ndGlann polarizer P2 and starts the switching part of the slicer via the photodiode PD. The transistor chain has been replacing a Krytron switch which has been found to produce too much noise in the electrical loop. The new transistor chain's switching level is arranged to switch just before the maximal pulse of the train, then the transistor chain gives the required quarter wave voltage on both Pockels cells, and the net effect of this operation is to rotate the plane of polarization of the linearly polarized light semi-train about 45°, to become circularly polarized one. And this is provided by passing the required pulses through the first pockel cell with quarter voltage applied on its electrodes. After passing the second pockel cell, the plane of polarization is again rotated about 45°. After the output pulse or semi-train of pulses exit from the second pockel cell to the second Glann polarizer, the quarter-wave voltage is removed from the pockel cells and the rest of the pulses from the train is again reflected by the Glann polarizer. The amplifier chain consisted of three Nd: YAG crystals with the diameter of 7 mm, and length of 120 mm. Two of them are placed in a single gold-plated double elliptical cavity and been pumped by a single xenon flashlamp. The rod of the last amplifier is placed in an elliptical silver coated cavity, and of its length, a 100 mm is being pumped by one xenon flash lamp. Before entering the last amplifier the laser beam is expanded twice by the telescope T_2. The amplification of the last amplifier is 10 times, resulting in 100 mJ output energy in the spot of diameter 5 mm and for input pumping energy for the last amplifier of 60 J. The oscillator output pulse's duration is further compressed due to the frequency doubling by the KDP nonlinear crystal to become 17 ps in the green. The power conversion efficiency has been found to be 50 %. The beam is expanded 2.5 times, and the output beam divergence could be adjusted through the range 0.1–1 milliradians according to which distance we want the beam to travel to the satellite, in order to obtain a reasonable and enough reflected signals from the satellite retro-reflectors as shown in Fig. 3. The laser pulses from the output telescope are directed by other four Code' mirrors to exit the mount in the direction of the satellite, and follow its orbit automatically with slight motor corrections during tracking, when required. These corrections could be easily determined by tracking a star near the same direction and adjusting the beam on it, then see the error and use it later with the satellite at or near the same direction of the star. The main units controlling this operation are the PC, and the LRE (Laser Ranging Electronics) which provide the laser system power supplies with triggers, and the Azimuth-Elevation steeper motors of the mount with the calculated orbital data (Ephemeris) as a function of time, which should be previously fed to the especially-prepared satellite ranging PC- programs. The saturation process in MQWs can be better quantified by the pulse fluence Φ than by the intensity I because of the limited relaxation time τ. To minimize the losses, the absorber should be saturable with the expected pulse fluence Φ. Another limitation is the damage threshold of the MQWs. A typical saturation fluence Φ_{sat} is about 70 μJ/cm². In the laser cavity the incident pulse fluence Φ can be adjusted by varying the illuminated area on the MQWs. If the intracavity pulse power is low, e.g. because of low pump power, then tighter focusing helps to achieve the necessary saturation fluence Φ_{sat} of typically some ten μJ/cm²[14]. The saturable absorption A of the MQWs can be calculated by:

$$A = \frac{A_0}{1 + \dfrac{\Phi}{\Phi_{sat}}}$$

(1)

where,

   Φ Pulse fluence (J/m²)
   A   Absorption
   Φ_{sat} Saturation fluence (J/cm²)
   A_0 Small signal absorption

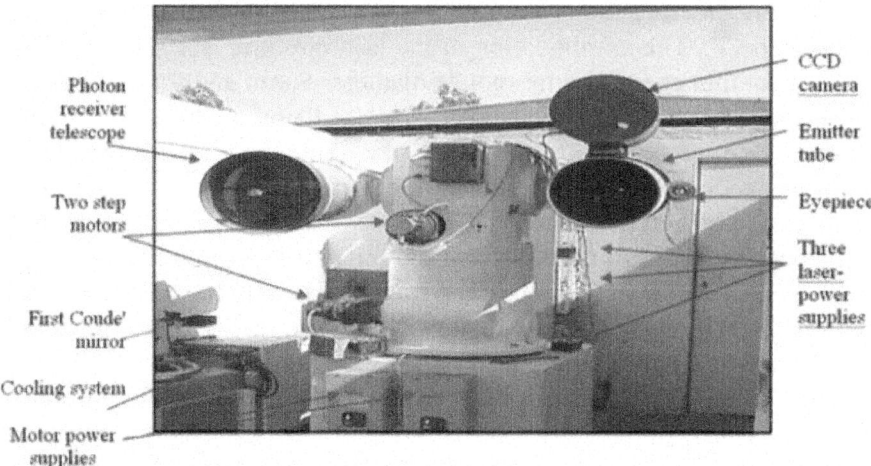

Figure 3: The outer part (Mount) of the automatic, Nd:YAG laser station

Figure 4 shows the saturable absorption A as a function of the fluence at constant $A_0 = 1\%$ and $\Phi_{sat} = 0.1$ mJ/cm². The pulse fluence Φ can be derived from the mean output power P as follows

$$\Phi = \frac{P}{(1 - R) \cdot f \cdot a}$$

(2)

where,

Φ Pulse fluence (measured in J/cm²)

P Mean output power of the laser

R Reflectance of the output mirror

f Repetition rate of the laser aIlluminated area on the MQWs

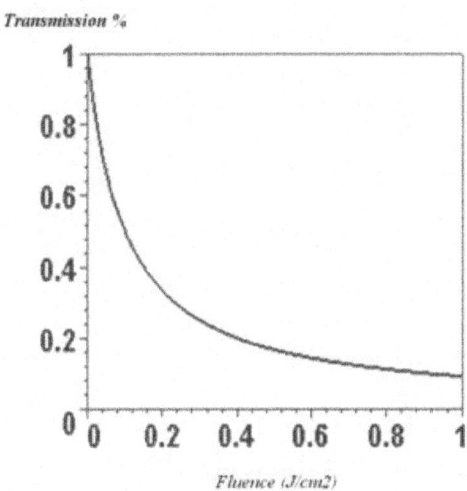

Transmission %

Fluence (J/cm2)

Figure 4: Shows the transmission as a function of the fluencyat constant A_0 = 1 % and Φ_{sat} = 0.1 mJ/cm²[14]

From Heisenberg's uncertainty principle for the conjugated variables pulse width Δt and photon energy $h\nu$. The TBWP (time bandwidth product) of a laser pulse is limited to about $\Delta t\, \Delta\nu \geq 1/\ (2\,\pi)$. Where h is a Planck's constant $(6.626\, x\ 10^{-34}\, J.s)$, ν is the pulse mean frequency and $\Delta\nu$ is the pulse bandwidth.An accurate calculation can show that the minimum TBWP for a Gaussian pulse is $\Delta t\, \Delta\nu \approx$ 0.44 (pulse duration in seconds multiplied by the pulse bandwidthin $Hz \geq 0.44$). The minimum TBWP for a Sech² pulse is $\Delta t\, \Delta\nu \approx 0.32$. Using the relation $c = \lambda\, \nu$ the frequency interval $\Delta\nu$ is related to the wavelength interval $\Delta\lambda$ by, $\Delta\nu = -c\, \Delta\lambda/\lambda^2$ Where, $c \approx 2.988\ x\ 10^8$ m/s is the speed of light in the vacuum.

For passive mode-locking by MQWs, the saturation of absorption for high incident intensity is essential. Therefore, nonlinear characteristics of the absorbers were studied practically using a similar multiple-quantum-well (MQWs) structures that were grown on 400 μm thick GaAs substrate at Center for High Technology Materials, University of New Mexico, by MBE method. They have 100 periods of $In_xGa_{1-x}As$ (x = 0.2) thickness of 8 nm, between two GaAs layers, each 10 nm thick. As a light source they used microchip diode pumped solid state laser Nanolase NP-02012-100 generating 810 ps long pulses with energy of 2.9 μJ at repetition rate of 15.3 kHz. The output radiation at average power of 44 mW was focused using a positive lens with focal distance of 50 mm. The insertion of the sample at various distance from the focal point enabled the change of the incident power density on the MQWs sample[15]. The output power from the laser was attenuated in order not to damage the MQWs; the maximum power density in the focal point was 10 MW/cm^2. The initial low power transmission of 23% far from the focal point increases to the 45% close to the focal point, the transmission of the GaAs plate alone was 52%, which means that non-saturable losses in MQWs were about 10% was mainly due to beam defocusing in GaAs[16]. The Nd:YAG laser system oscillator at Helwan SLR is used to test the three MQWs semiconductor saturable absorbers as shown in Fig. 5. Each MQW sample is installed independently at about 45⁰ with the resonator axis and the dye cell is filled only with dichloroethan in the presence of MQWs. The full output (mode-locked) train from the oscillator is directed via the system laser mirrors to a photodiode which is connected to oscilloscope.

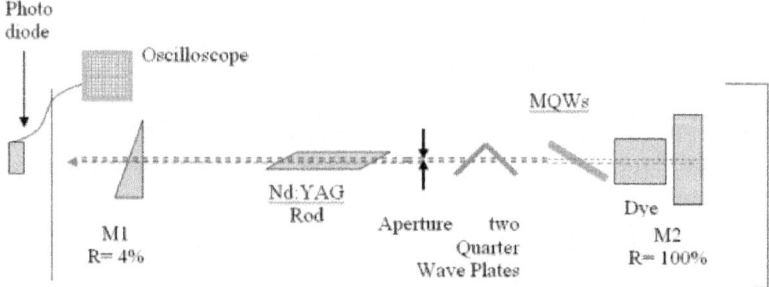

Figure 5: Setup for MQW saturable absorber, mode-locking

RESULTS AND DISCUSSIONS

For each dye solution the energetic stability of the output burst from the oscillator was measured. In the first case only passive and in the second case passive and active switch of the Q-factor was used. Fig. 6 and Fig. 7 showthe bursts of output pulses. Fig. 6shows the passive modulation of the resonator (saturable absorber- dye 3955 in ethylalcohol) while Fig. 7 shows the passive (with the same saturable absorber) and also active modulation. An energy stability and also general laser stability is perceptible. Each flash of the flash lamp is followed by laser action (there is no pulse drop out). Fig. 8 and Fig. 9 show the same bursts in 50-multiple exposition, which also demonstrates the oscillator stability. Fig. 10a – f. show a statistic division as well as histograms. A measured energy is placed on horizontal axis and normalized percent of occurrence on vertical axis. Then each histogram is completed with percentage dispersion of observed quantity for the given oscillator configuration.

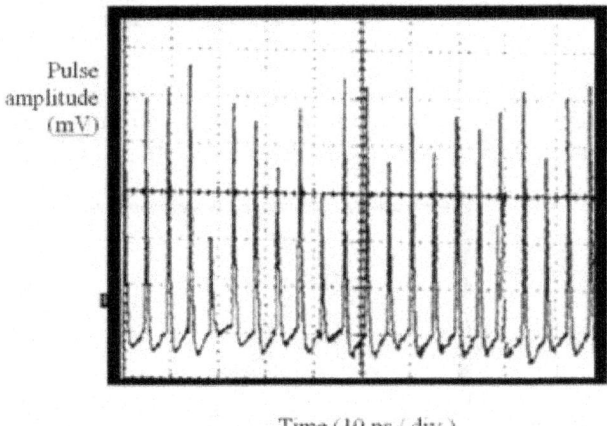

Time (10 ns / div.)

Figure 6: The energetic stability of the oscillator output, 21 successive pulses, passive modulation- dye 3955 in ethylalcohol, vertical scale: 5 mV/ div

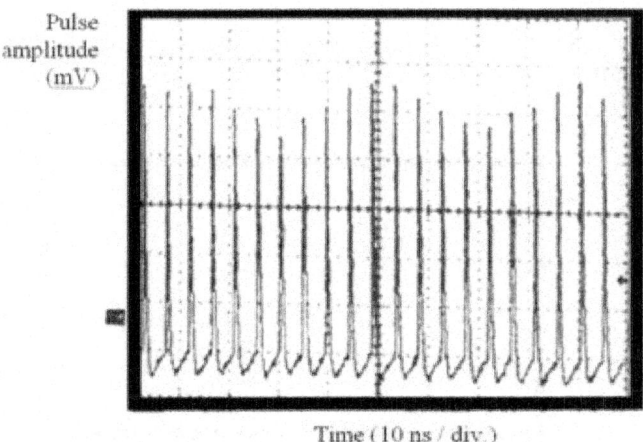

Time (10 ns / div.)

Figure 7: The energetic stability of the oscillator output, 21 successive pulses, passive (dye 3955 in ethylalcohol) + active modulation, vertical scale: 5 mV/ div

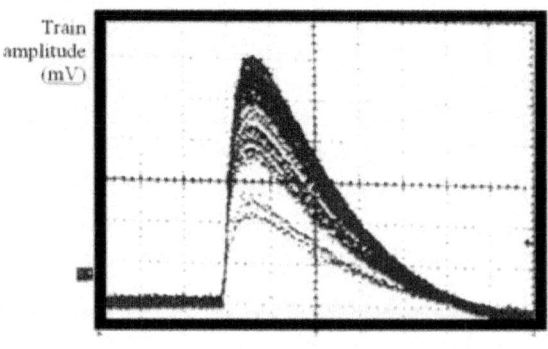

Figure 8: Energetic stability of oscillator output – 50 multiple exposions. passive modulation – dye 3955 in ethylalcohol,vertical scale:5 Mv/ div

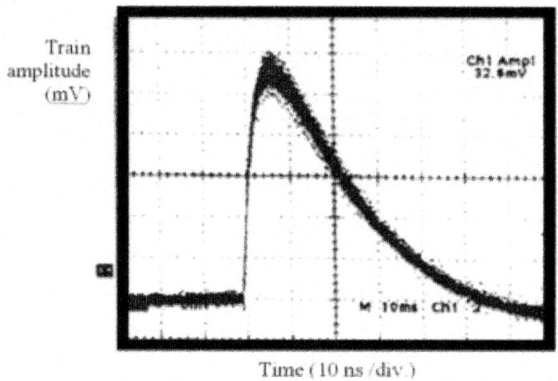

Figure 9: Energetic stability of oscillator output - 50 multiple exposions. passive modulation – dye 3955 in ethylalcohol + active modulation,vertical scale: 5 mV/ div

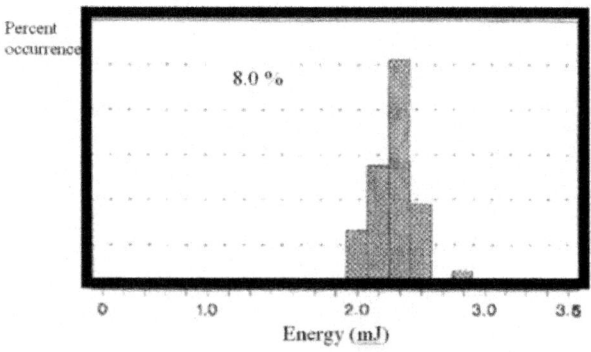

(a) Passive switching (3955 in ethylalcohol)

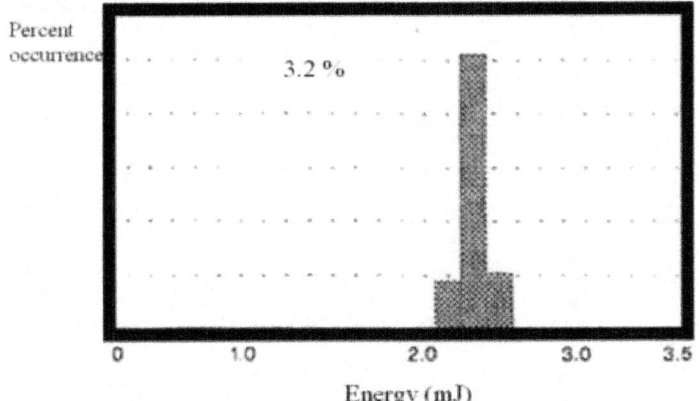

(b) Passive switching (3955 in ethylalcohol) + active switching

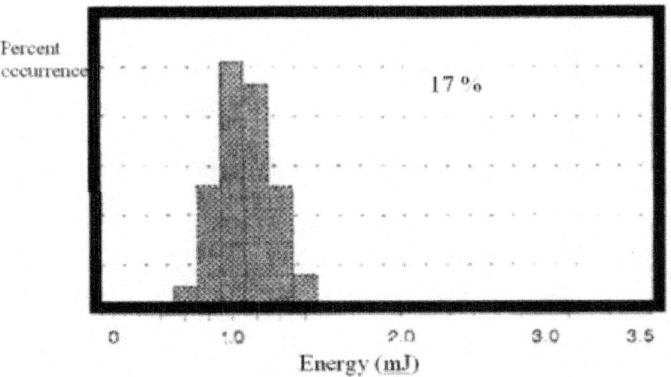

(c) Passive switching (ML63 in ethylalcohol)

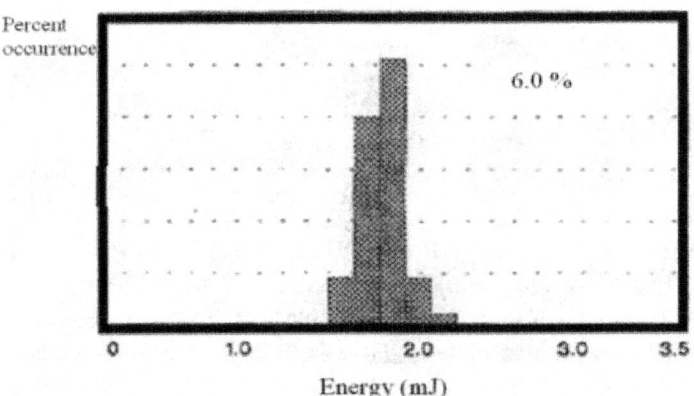

(d) Passive switching (ML63 in ethylalcohol) + active switch

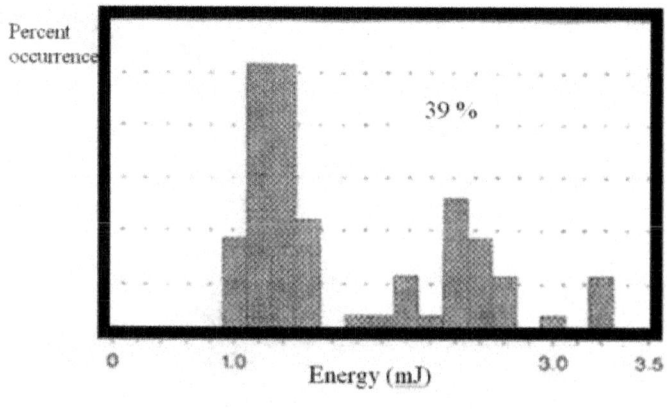

(e) Passive switching (ML51 in dichloride)

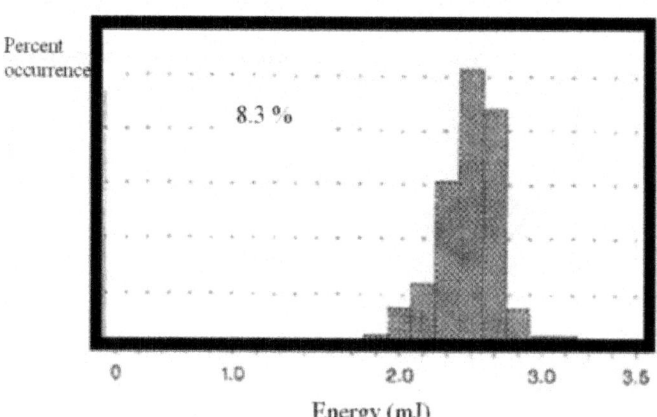

(f) Passive switching (ML51 in dichloride) + active switching

Figure 10: Histograms of bursts energy distribution

Oscillator Output

The output from the oscillator is trains of TEM_{00} mode-locked pulses of wavelength 1064 nm and of 24 ps FWHM measured by streak camera, its repetition rate is 5 Hz and having energy of about 1 mJ in single pulses, about 8 mJ per train of pulses, the pulses are separated by the roundtrip time interval of 6 ns (the length of the oscillator is 0.90 m), the time resolution of the streak camera was 1.4 ps. However, the oscillator output could vary both quantitatively (number of pulses per train) and qualitatively (energy and time duration of pulses as well as the envelope of the train) the reason is the flash lamp input energy and the type of Q-switch, the Q-switch is a liquid dye. Also various experimental parameters such as the type of dye, its initial transmission, and its circulation speed inside the cell play important roles.

Output Energy Stability

The amplifiers in laser cavity affect the stability of output parameters only if they are in saturation and thus fluctuations are restricted higher from the certain level of the signal. The optimum working dye (ML51) - circulation voltage and the stability of the mode-locked pulse energy have been measured, and new mode-lockers have been investigated.

Optimum Dye-circulation Voltage

The dye-solution must circulate inside the cell prior to and during the laser system operation in order to keep working fast and properly, for this reason a simple experiment was done to measure the optimum circulation voltage for the dye ML51 of anonymous initial transmission, T. The Photodiode and the station calibration program can calculate the number of received laser shots by the photo detector and the number of applied external laser triggers to the flash lamp. The dye circulation inside the cell at Egyptian Helwan station is accomplished by a small motor of variable speed, fixed in a vertical position with a small magnet fixed with its rotating axis. Both are located in a cavity under the dye cell, with its rotating magnet up and just below the dye cell, another small cylindrical magnet is located inside the dye cell. When the motor is on, the magnet connected to the motor axis rotates under the dye cell and, therefore making the other magnet inside the dye cell to rotate in the same direction, hence circulating the dye inside the cell. The efficiency of the dye, clearly, depends on its circulation speed, because when the dye circulation was too slow (applied voltage < 6 volts) many laser shots were lost from the output with the appearance of noise mainly due to incomplete mode-locking. Figure 11 shows the average number of successful laser shots per 1000 triggers versus the applied voltage to the mixing motor. On the other hand, when the voltage was higher than 12 Volts, i.e., the dye circulating was very fast, there had been a sudden output drop to below 15% between 12 and 13 volts, and then rises again to above 40% near 14 Volts, and after that it decreases fast to reach 10% output at near 16 volts. It is foundthat the optimum circulation speed of the dye was at 11-12 volts range as input for the rotating motor, as the output complete-laser shots per 1000 triggers reached 99% only in this range. Nevertheless, it could circulate with a fairly well output in the range of 6-12 voltswith a maximum loss of 3% only as seen from the behavior of the plateau in Fig. 11.

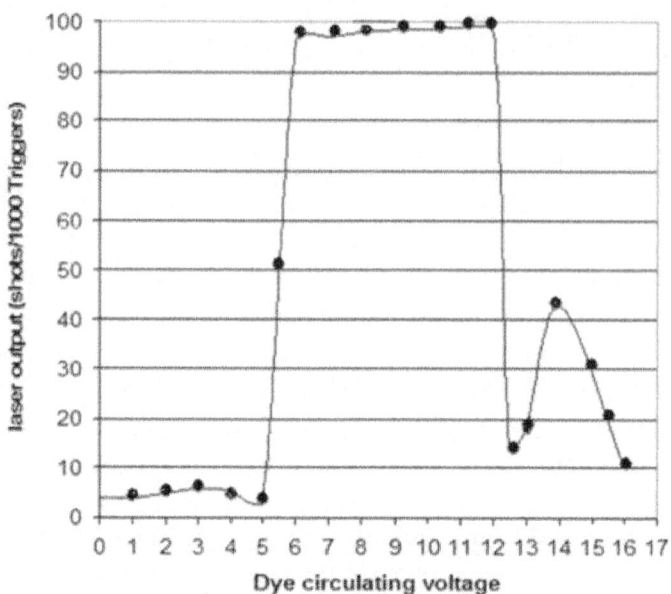

Figure 11: The average number of successful laser shots per 1000 triggers versus the applied voltage to the motor

Effect of Dye Transmission and Input Electrical Energy on the Output Energy

The average laser output energy E_{out}(av.) is calculated for every 10 readings of the energy meter taken for each dye transmission. The same procedure was made for 7 different input electrical energy values. The results are plotted in Fig. 12 versus the flash lamp input electrical energy E_{in} for three different transmissions (25%, 30%, and 40%) of the dye ML51 dissolved in dichlorethan.

Figure 12 shows the output laser energy shows a nonlinear behavior with the input flash lamp energy at lower initial dye transmission 25%. As the dye initial transmission increases 30% and 40% however the change appears to be more linear. It is also noticed that all the used dye transmissions, result in the same output energies at two flash lamp input energies, and this is clear from the two intersection points between the three curves representing the three dye transmissions together at around (80 J, 45 mJ), and near to (102 J, 115 mJ). The lower dye transmission, 25%, gives, in average the least value of output energy, then that of 40%, and the greater average value is obtained from the 30% dye transmission for the dye ML51 used in the selected range from 70 to 109 J of the input flashlamp energy.

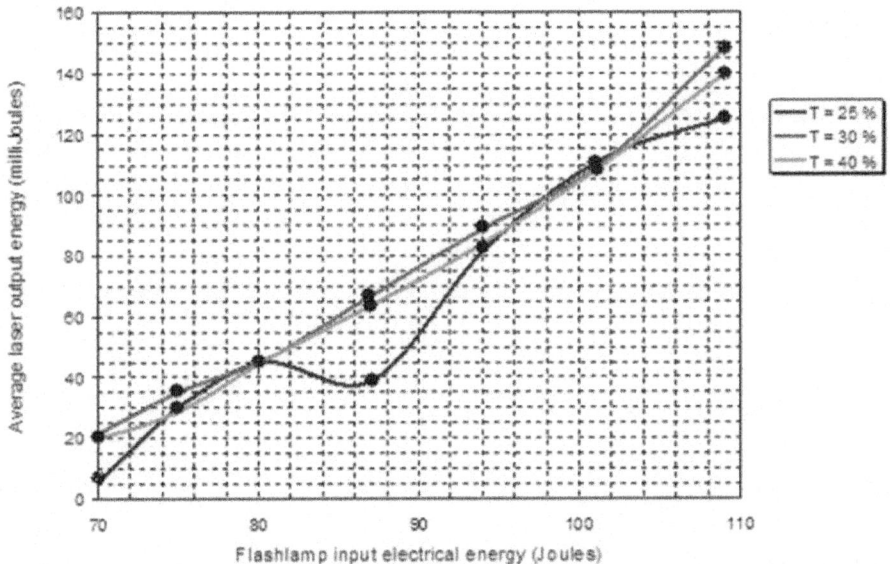

Figure 12: The change of the output laser energy per train with the input pumping energy for three different initial dye transmissions; 25 %(the blue line), 30% (the green line), and 40 % (the red line)

Furthermore the curve at 25% transmission suffers from big variations in the output energy in the ranges of input energy 70-74 J; 80-94 J; and also 104 to above 108 J, however the output energy for this dye transmission only exceeds that from the other two in the range of 98-102 J of flashlamp input energy. The other two curves for the dye transmissions 30%, and 40% shows greater average output energy and much better output energy stability. From these experiments, columns A, B, and C as shown in Fig. 13, three average train samples from each sample of the MQWs 758, 1487, and 2313 respectively are obtained. It can be seen that the MQWs sample No.758 (column A) gives mode-locked trains of reasonable symmetry and greater amplitude (energy), (peak pulse: 1.2 to 1.4 volts on the oscilloscope scale) than the other two MQWs (columns B and C) which results in mode-locked pulse trains of less symmetry and of less peak voltage. The resulting mode-locked trains from liquid dye ML51(A), and from MQW number 758 saturable absorber (B), are compared as shown in Fig. 14. It is found that the dye ML51 mode-locking gives an average peak pulse of about twice that resulting from MQW sample No.758.

Column A: B: C:

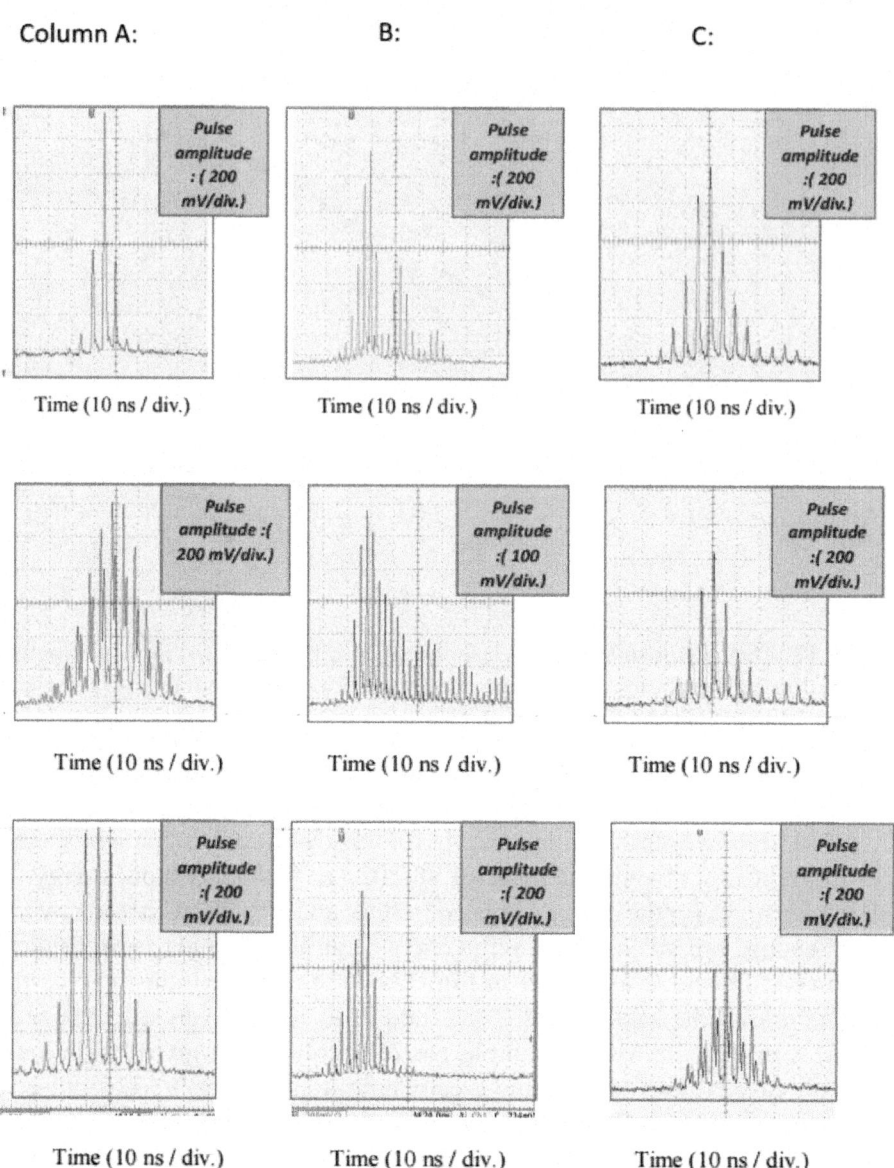

Figure 13: Mode-locked train results when: (A), MQW No. 758; (B), MQW No. 1487; (C), MQW No. 2313 were introduced inside the Helwan SLR laser system-resonator as a passive mode-locker

A

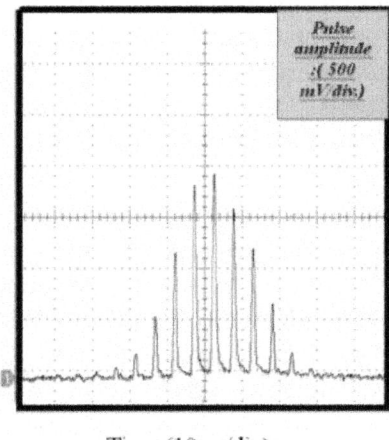

Time (10 ns/div)

B

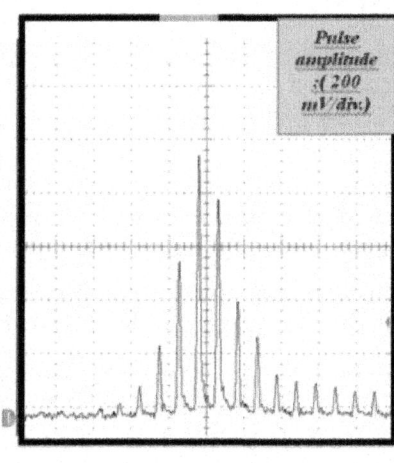

Time (10 ns/div)

Figure 14: Mode-locked train when: (A) dye ML51, and (B) MQW No.758 were used

CONCLUSIONS

The goal of this work was to develop the construction of the Helwan SLR laser system in Egypt, to measure its output energy stability from the saturable absorber dye ML51 at three different concentrations in dichlorethan and from different pumping energies, and at the optimum dye circulation speed. The research measurement of Q-switches here was made on the bases of the laser radar. Double switching experiments: showed that the best result was reached when the acousto-optic modulator and the dye solution 3955 in ethylalcohol (percent occurrence 3.2%) were used together. From the statistic division of the measured bursts energies, one can conclude that the resonator stability is increased by factor of about 2.5 times, when dual switch of the Q-factor is used.It can be seen that the MQW No.758 gives mode-locked trains of reasonable symmetry and greater energy, (peak pulse: 1.2 to 1.4 volts) than the other two MQWs. Furthermore, the dye ML51 gives an average peak of about 1.8 volts while the MQW No.758 gives a train of peak pulse less than 1 volt and of less symmetrical than that of the dye ML51 mode-locker. Therefore, the organic dye ML51 is better than the semiconductor MQW saturable absorbers for mode-locking HelwanNd:YAG laser system for SLR purpose because the small energy in the train means insufficient power of the laser beam and therefore inability to track even the low orbit satellites especially at lower elevations because of the high atmospheric losses.

ACKNOWLEDGMENTS

Authors would like to thank Prof. Dr. A. Novotny for fruitful discussions.

REFERENCES

1. A. A. I. Khaliland M. C. Richardson, "Generation of ion beams on a steel surface by laser radiation at various wavelengths under the presence of an applied electric field". Laser Phys Lett.3(3), 137-144 (2006).

2. A. A. I. Khalil, "Spectroscopic studies of UV lead plasmas produced by single and double pulse laser excitation". Laser Phys. 23,015701(2013).

3. Ji-ying Peng, Bao-shan Wang, Yong-gang Wang, Jie-guang Miao, Hui-ming Tan, Long-sheng Qian and Xiao-yu Ma, "Q-switched and mode-locked diode-pumped Nd: YVO4 laser with an intracavity composite semiconductor saturable absorber". Optics & Laser Tech. 40(2), 243-246 (2008).

4. G. Xiao and M. Bass, "A generalized model for passively Q-switched laser including excited state absorption in the saturable absorber". IEEE J Quantum Electron, 33 (1), 41–44 (1997).

5. H. Jelínková, P. Cerny, J. Šulca, J. K. Jabczynski, K. Kopczynski, Z. Mierczyk, M. Miyagi, Y. Matsuura, and Y.-W. Shi,»Nd: YAP 1.34 um/1.08 um laser passively mode-locked and Q-switched by V3+:YAG/ BDN II saturable absorbers with efficient radiation delivery through a

hollow glass waveguide coated with COP/Ag». Opt Eng.41(8), 1976-1982 (2002).

6. J. K.Jabczynski, W. Zendzian, J. Kwiatkowski, V. Kubecek, H. Jelinkova, A. Stintz and J. C. Diels,»Picosecond Nd:YAG slab laser passively Q-switched and mode-locked, using multiple quantum well saturable absorbers». SPIE, 6731, 67312I (2007).

7. Y.Tang, A. Siahmakoun, G. Sergio, M. Guina and M. Pessa, «Opticalswitching in a resonant Fabry–Perot saturable absorber». J Opt A: Pure Appl. Opt.8, 992 (2006).

8. V. Kubecek, J. Biegert, J-C Diels and M. R.Kokta,»Practical source of 50 ps pulses using a flashlamp pumped Nd:YAG laser and passive all-solid state pulse control». Opt Commun. 177(1-6): 317-321 (2000).

9. U. Keller, «Ultrafast all-solid-state laser technology». Appl. Phys. B, 58, 347-363 (1994).

10. J. Biegert, J-C. Diels, K. Malloy, A. Stintz, V. Kubecek, «42 ps pulses from a flashlamp pumped Nd:YAG laser using all-solid-state passive pulse control».Conference Digest, Conference on Lasers and Electro-Optics Europe, IEEE Cat. No. 00TH8505, 115 (2000).

11. A. M. Malyarevich, V. G. Savitski, P. V. Prokodshin, N. N. Posnov, K. V. Yumashev, E. Raben, A. A. Zhilin,»Glass doped with PbS quantum dots as a saturable absorber for 1-m neodymiumLasers». JOSA B, 19,28-31 (2002).

12. A. A. I. Khalil, «A comparative spectroscopic study of single and dual pulse laser produced UV tin plasmas». Opt. & Laser Tech. 45, 443-452 (2013).

13. V. Kubecek, J. Biegert, J-C.Diels, A. Dombrovsky, K. Malloy and A. Stintz,»Picosecond flashlamp pumped Nd:YAG laser using all-solid-state passive pulse control». Proc. of SPIE, 4424, 151-154 (2001).

14. Batop company of optoelectronic devices,http://www.batop.com

15. R. Schwedler, H. Mikkelsen, K. Wolter, D. Laschet, J. Hergeth and H. Kurz, Institut fur Halbleitertechnik, RWTH Aachen Sommerfeldstrasse 24, 52074 Aachen, Germany, «In GaAs/InP multiple quantum well modulators in experiment and theory».J. Phys. III France,2341 (1994).

16. V. Kubecek, H. Jelínková, A. Dombrovský,A. Stintz, J. Diels. «Flashlamp pumped oscillator-amplifier Nd: YAG system mode locked using multiple quantum well saturable absorber». presented at Solid State Lasers conference 5460-15, part of Photonics Europe, Strasbourg, 26-30, Technical program, 195 (2004).

Chapter 14

OPEN STATISTICAL ISSUES IN PARTICLE PHYSICS

Louis Lyons

Oxford University

ABSTRACT

Many statistical issues arise in the analysis of Particle Physics experiments. We give a brief introduction to Particle Physics, before describing the techniques used by Particle Physicists for dealing with statistical problems, and also some of the open statistical questions.

INTRODUCTION

Particle Physics tries to delve into the structure of matter at its most basic level. It continues a tradition that dates back to the Greeks2 or even earlier. In the early days of Chemistry, the smallest entities were atoms. Early in the 20th century, the experiments of Rutherford demonstrated that atoms consisted of a small nucleus, with the electrons circulating at distances of ~ 10−10 metres. Subsequently, the nucleus was found to be made of protons and neutrons. Many other particles (known as hadrons) like protons and neutrons have subsequently been discovered, but within the last 30 years, the quark model has brought understanding to the multitude of what used to be called "elementary particles." The entities that we currently believe are fundamental (i.e., they do not seem to have any sub-structure) are the quarks and leptons shown in Table 1. There are 6 of each, and they appear to be arranged in 3 "generations" of increasing mass, each containing quarks of electric charge +2/3 and −1/3 (in units where the electron's charge is −1) and leptons of charge −1 and 0. The neutral leptons are called neutrinos. Although charged leptons and neutrinos have been detected, quarks are believed to be confined within hadrons. They have not been observed directly, but their existence is inferred from the simplification they bring to the multitude of hadrons, and to the way they explain many features of the way hadrons interact with each other or with leptons.

Table 1: The basic particles

Particle, charge	Generations		
	1	2	3
Quark, +2/3	u (0.3)	c (1.5)	t (175)
Quark, −1/3	d (0.3)	s (0.5)	b (5)
Neutrino, 0	$\nu_e(< 3*10^{-9})$	$\nu_\mu(< 2*10^{-4})$	$\nu_\tau(< 0.02)$
Lepton, −1	electron $(5*10^{-4})$	$\mu(0.1)$	$\tau(1.8)$

Masses shown in brackets are in GeV/c². In these units, the mass of the proton is 0.9.

In addition to these particles, there are also others responsible for mediating the various fundamental forces. These include the massless photon γ, responsible for the electro-magnetic force; the massive W and Z bosons which mediate the weak force; and the gluons g responsible for the strong force. In addition, there is the still to be detected graviton which mediates gravitational forces, and is usually denoted by the symbol. Because the interacts so weakly it is hard to observe. Finally, there is the undiscovered Higgs boson, which is believed to be responsible for the mass of the other particles, and which is the object of intense searches in current experiments. Of course, theoretical physicists are prolific at inventing models, and so there are many other suggested particles.

Experiments in Particle Physics are usually conducted at large accelerators, for example, at the European Centre for Nuclear Research (CERN) in Geneva, or at Fermi National Accelerator Lab near Chicago. CERN's soon-to-be-running Large Hadron Collider (LHC) is in a tunnel about 100 metres below the surface and 27 kilometres in circumference, and which straddles the French–Swiss border. Protons circulate in bunches in opposite directions around the ring, and collide with each other at the center of large detectors. The bunches are about the width of a human hair, and are ~10 centimetres long. When they collide, new particles are produced by converting the available kinetic energy into mass. The detectors are designed to track the path of each particle, measure its curvature in the magnetic field and hence determine the particle's momentum, and also to give information on the particle's identity (e.g., whether it is an electron, muon, pion, kaon or proton). Reactions between colliding protons will occur at a very high rate, but most of them are fairly uninteresting. Thus, experiments are designed to have a trigger, which makes a very fast decision as to whether the collision (called an "event") is likely to be interesting, and hence whether the data from the detector is worth storing. Because of data read-out and storage constraints, only about 100 events per second are recorded, and each may contain about a Megabyte of information. Since the accelerator may run for 15 years, some 10^{10} events can be collected by each experiment. In analyzing data, allowance must be made for the distorting effect introduced by any selection bias of the trigger.

This review attempts to present some interesting statistical issues in the analysis of data collected in Particle Physics experiments. The items discussed below are a mixture of current practice, ideals to which we aspire and some personal prejudices of the author. It is hoped that the approaches mentioned in this article will be interesting or outrageous enough to provoke some Statisticians either to collaborate with Particle Physicists, or to provide them with suggestions for improving their analyses. It is to be noted that the techniques described are simply those used by Particle Physicists; no claim is made that they are necessarily optimal.

A Glossary of Particle Physics terminology appears in the supplementary material [Lyons (2008)].

PARTICLE PHYSICS ANALYSES

This section starts with two typical examples of Particle Physics analyses, the first involving parameter determination, while the second tests whether data is consistent with a null hypothesis H_0, or whether an alternative hypothesis H_1 is favored. Further examples are described later. More detailed descriptions can be found in the various papers of the PHYSTAT series of Conferences [see James, Lyons and Perrin (2000), Cheung and Lyons (2000), Whalley and Lyons (2002), Lyons, Mount and Reitmeyer (2003), Lyons and Unel (" 2005), Reid, Linnemann and Lyons (2006), Prosper, Lyons and De Roeck (2007)]. In particular, at the PHYSTAT-LHC meeting at CERN in 2007, the major experiments at the LHC presented their statistical "wish-lists" [Gross (2007), Belikov (2007), Xie (2007)].

Lifetimes

Here we estimate the lifetime of some specific particle. Thus, we could have n independent observations $t_1 \ldots t_i \ldots t_n$ for the times between the production and decay for this particle in the selected events. Then the mean lifetime τ could be determined by an unbinned likelihood fit to the probability density $\tau^{-1} \exp(-t/\tau)$. In real life we would have a more complicated expression, to allow for a possible background with a different time dependence, experimental resolution on the determination of ti , and experimental acceptance of the detector and the trigger, which depends on t.

The various steps in the data analysis include:

- Reconstruct tracks from the hits in the detector.
- Select wanted events that are enriched in the particle whose lifetime we wish to measure.
- For each interaction, extract the decay time t from L and v, the distance the particle travels and its speed. Typical values are picoseconds, mms and 99% of the speed of light respectively.
- Model the signal, typically by an exponential time dependence probably smeared by time resolution effects, and the background. Time-dependent efficiencies for collecting the data may also be relevant.
- Perform a likelihood fit, to determine τ and its statistical error σstat.
- Estimate the systematic error σ_{syst}, and quote the result as $\tau \pm \sigma_{stat} \pm \sigma_{syst}$. These systematics [Heinrich and Lyons (2007)] can arise from uncertainties in some of the extra parameters involved in modeling the data (e.g., the level of background contaminating our signal), or from possible uncertainties in the theory (maybe the expected exponential decay distribution is complicated by the existence of two overlapping particles). Statisticians usually refer to the former as "nuisance parameters." In analyses involving enough data to achieve reasonable statistical accuracy, considerably more effort is devoted to assessing the systematic error than to determining the parameter of interest and its statistical error.
- Assess the goodness-of-fit between the data and the model, and ignore the estimated value for the parameter if the fit is unsatisfactory.

Significant Peak

Another type of analysis might consist of looking at a mass spectrum (see Figure 1). In many situations we would expect to observe a rather smooth and somewhat boring distribution, but sometimes there may be a significant-looking peak at some mass position. This could correspond to the exciting discovery of a new particle, to a boring statistical fluctuation of the smooth background or to some unfortunately overlooked effect in the analysis.

We can make some numerical statement about the probability of obtaining a statistical fluctuation at least as extreme as the one we have observed. In this situation, we are performing a "Goodness of Fit" test, that is, we are comparing our data with the null hypothesis of a smooth distribution.

Alternatively and probably more sensitively, we could use our data to compare the two hypotheses—just smooth background or an interesting peak above the background; this is "Hypothesis Testing."

Bayes or Frequentism

In many analyses the question arises whether to use a Bayesian or a Neyman–Pearson Frequentist approach, or one which is neither (e.g., χ^2, likelihood, etc.). Particle Physicists tend to favor a frequentist method. This is because we really do consider that our data are representative as samples drawn according to the model we are using (decay time distributions often are exponential; the counts in repeated time intervals do follow a Poisson distribution, etc.), and hence we want to use a statistical approach that allows the data "to speak for themselves," rather than our analysis being dominated by our assumptions and beliefs, as embodied in Bayesian priors. The reluctance to use priors is strongest in situations with several variables where multidimensional priors would be required, or in cases where very little is known about the relevant parameter—it may be acceptable to use prior information about a parameter which is already well measured, but more problematic to try to quantify prior ignorance.

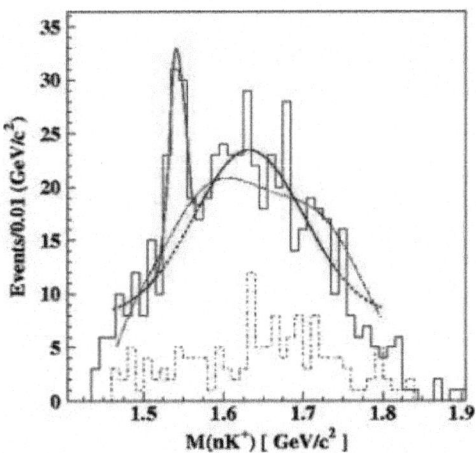

Figure 1: Mass histogram. This is for reactions producing a neutron (n), π^+, K^+ and K^-. A histogram of the effective mass of the nK$^+$ combination is plotted. If a particle decaying into a neutron and a K+ is produced in these reactions, a narrow peak should appear in this histogram at the particle's mass, but if not the distribution should be smooth. The curve is an attempt to deduce this smooth background. Does the histogram provide evidence for a new particle, as opposed to there being a statistical fluctuation from the smooth background, and/or an incorrectly estimated

background? A new particle here would be very interesting, as it would not fit into the simple quark model, because it would require a 5-quark structure.

However, in practice, it is very hard to use the Neyman frequentist construction when more than two or three parameters are involved: software to perform a Neyman construction efficiently in several dimensions would be most welcome. The choice of a useful ordering rule is also very important. Thus from a pragmatic point of view, even ardent frequentists are prepared to use Bayesian techniques. Most of them, however, would like to ensure that the technique they use provides parameter intervals with reasonable frequentist coverage. There are even mixed methods [Cousins and Highland (1992)] that use Bayesian priors for nuisance parameters, but a frequentist method for the parameter of interest. The thinking here is that, although such an approach cannot be justified from fundamentals, it provides a practical method whose properties can be checked, and are often satisfactory.

Particle Physicists would appreciate advice on how to construct priors for parameters of interest, to be used in conjunction with information-based priors for nuisance parameters, and which might give reasonable coverage [see Demortier (2005)].

EXPERIMENTAL DESIGN

Because experimental detectors are so expensive to construct, the time-scale over which they are built and operated is so long, and they have to operate under harsh radiation conditions, great care is devoted to their design and construction. This differs from the traditional statistical approach for the design of agricultural tests of different fertilisers, but instead starts with a list of physics issues which the experiment hopes to address. The idea is to design a detector which will provide answers to the physics questions, subject to the constraints imposed by the cost of the planned detectors, their physical and mechanical limitations, and perhaps also the limited available space. Inevitably, compromises in the design are required, and testing of any proposed scheme involves analysis of the simulated "data" to see if the physics aims can indeed be achieved.

Design is also involved when planning what technique is to be used to analyze the experiment's real data. This will be especially detailed if a blind analysis is to be performed (see Section 8).

Another example is provided by the attempt to assess the systematic error on an estimated parameter, caused by nuisance parameters. This often requires producing simulations of the data with different values of the nuisance parameter, and seeing how much the physics parameter's value changes when the nuisance parameter value is changed by its uncertainty (compare Sections 5.4 and 6.2 for ways of incorporating nuisance parameters in upper limit and in p-value calculations respectively). When several nuisance parameters are involved, there is the question of whether separate simulations should be produced, in each of which only one of the nuisance parameters is changed from its optimal value by its uncertainty; or whether it is better to generate simulations in each of which all nuisance parameters are simultaneously changed from their optimal values according to their expected (possibly correlated) multivariate distribution. The two methods are sometimes referred to unisim (or OFAT = One Factor At a Time) and multisim respectively. The question is which method requires less computing time to achieve the same accuracy for the systematic error [Roe (2007)]. How to assess systematics was much discussed at the Banff meeting [Reid, Linnemann and Lyons (2006)] and PHYSTAT-LHC [Read (2007), Neal (2007), Linnemann (2007)].

SEPARATING SIGNAL FROM BACKGROUND

Almost every Particle Physics analysis uses some technique for separating signal from background. This is because only a fraction f of the complete set of stored events (which because of the trigger can be a factor of 10^7 down on the total reaction rate) will contain interactions of interest for the analysis being performed. Depending on the investigation being undertaken, f could be as small as 10^{-8}.

First some simple "cuts" are applied; these are generally loose selections on single variables, which are designed to remove background while barely reducing the signal. For example, the selected events could be required to have no more than 6 charged tracks. Then some more sophisticated analysis is performed, perhaps using more complicated derived variables, for example, the mass of a possible particle decaying into a kaon and 3 charged pions. To separate signal from background in the multi-dimensional space of the event observables, these analyses typically use methods like Fisher discriminants, boosted decision trees, artificial neural networks (including Bayesian nets), support vector machines, etc. [Prosper (2002), Friedman (2003, 2005)]. A description of the software available for implementing some of these techniques can be found in Narsky (2006) and H"ocker (2007).

If a large data sample is available to perform an accurate measurement of a property of some particle, then it is not a disaster if there is some level of background in the finally selected events, provided that it can be accurately assessed and allowed for in the subsequent analysis. At the other extreme, the separation technique may be used to see if there is any evidence for the existence of some hypothesised particle (the potential signal), in the presence of background from well-known sources. Then the actual data may in fact contain no observable signal.

These techniques are usually "taught" to recognize signal and background by being given examples consisting of large numbers of events of each type. These may be produced by Monte Carlo simulation, but then there is a problem of trying to verify that the simulation is a sufficiently accurate representation of reality. It is better to use real data, but the difficulty then is to obtain sufficiently pure samples of background and signal. Indeed, for the search for a new particle, true data examples do not exist. However, it is the accurate representation of background that is likely to pose a more serious problem.

The way that, for example, neural networks are trained is to present the software with approximately equal numbers of signal and background events[3], and then to optimize the cost function C for the network. This is defined as $C = \Sigma(z_i - t_i)^2$, where z_i is the trained network's output for the ith event; t_i is the target output, usually chosen as 1 for signal and zero for background; and the summation is over all testing events presented to the network. The problem with this is that C is not what we really want to optimize. For a search for a new particle, this could be the sensitivity of the experimental upper limit in the absence of signal, while for an analysis measuring the properties (such as mass or lifetime) of some well-established particle, we would be interested in minimizing the error (including systematic effects) on the result.

So the open questions are as follows:

Is it possible to define what multivariate method will perform well in a given class of problems?

- How can we check that our multi-dimensional training samples for signal and background are reliable descriptions of reality?
- How many events are required for training?
- How should they be divided between signal and background, especially when there are several different sources of background?
- What is the best way of allowing for nuisance parameters in the models of the signal and/or background?
- Are there easy ways of optimizing on what is really of interest?

UPPER LIMITS

Most searches for new phenomena have not found any evidence for exciting new physics. Recent examples from Particle Physics include searches for the Higgs boson, supersymmetric particles, dark matter, etc.; attempts to find substructure of quarks or leptons; looking for extra spatial dimensions;

measuring the mass of a neutrino; etc. Rather than just saying that nothing was found, it is more useful to quote an upper limit on the sought-for effect, as this could be useful in ruling out some theories. An example of this was the experiment by Michelson and Morley in 1887 which attempted to measure the speed of the Earth with respect to the aether. No effect was seen, but the experiment was sensitive enough to lead to the demise of the aether theory

A simple scenario is a counting experiment where a background b is expected from conventional sources, together with the possibility of an interesting signal s. The number of counts n observed is expected to be Poisson distributed with a mean $\mu = \varepsilon * s + b$, where ε is a factor for converting the basic physics parameter s into the number of signal events expected in our particular experiment; it thus allows for experimental inefficiency, the experiment's running time, etc. Then given a value of n which is comparable to the expected background, what can we say about s? The true value of s is constrained to be non-negative. The problem is interesting enough if b and ε are known exactly; it becomes more complicated when only estimates with uncertainties σ_b and σ_ε are available.

Even without the nuisance parameters, a variety of methods is available. These include likelihood, χ^2, Bayesian with various priors for s, frequentist Neyman constructions with a variety of ordering rules for n, and various ad hoc approaches. The methods give different upper limits for the same data.4 A comparison of several methods can be found in Narsky (2000). The largest discrepancies arise when the observed n is less than the expected background b, presumably because of a downward statistical fluctuation. The following different behaviors of the upper limit (when n < b) can be obtained:

- Frequentist methods can give empty intervals for s, that is, there are no values of s for which the data are likely. Particle Physicists tend to be unhappy when their years of work result in an empty interval for the parameter of interest, and it is little consolation to hear that frequentist statisticians are satisfied with this feature, as it does not lead to undercoverage.

When n is not quite small enough to result in an empty interval, the upper limit might be very small. This could confuse people into thinking that the experiment was much more sensitive than it really was.

- The Feldman–Cousins frequentist method [Feldman and Cousins (1998)] that employs a likelihood-ratio ordering rule gives upper limits which decrease as n gets smaller at constant b. (This can also occur in other frequentist methods.) A related effect is the growth of the limit as b decreases at constant n. Thus, if no events are observed (n = 0), the upper limit for a 90% interval is 1.08 for b = 3.0, but 2.44 for b = 0. This is sometimes presented as a paradox, in that if a bright graduate student worked hard and discovered how to eliminate the expected background, they would be "rewarded" by obtaining a weaker upper limit.6 An answer is that although the actual limit had increased, the sensitivity of the experiment with the smaller background was better. There are other situations—for example, various random choices of measuring instruments [Cox (1958)]—where a measurement with better sensitivity can on occasion give a less-precise result.

- In the Bayesian approach, the dependence of the limit on b is weaker. Indeed, when n = 0, the limit does not depend on b.

- Sen et al. (2008) consider a related problem, of a physical non-negative parameter λ producing a measurement x, which is distributed about λ as a Gaussian of variance σ^2. As the observable x becomes more and more negative, the upper limit on λ increases, because it is deduced that σ must in fact be larger than its originally quoted value.

In trying to assess which of the methods is best, one first needs a list of desirable properties. These include:

- Coverage: Even most Bayesian Particle Physicists would like the coverage of their intervals to match their quoted credibility, at least approximately. Because the data in counting experiments are discrete, it is impossible in any sensible way to achieve exact coverage for all μ. However, it is not completely obvious that even Frequentists need coverage for every possible value of μ, since different experiments will have different values of b and of ε. Thus, even for a constant value of the physical parameter s, different experiments will have different $\mu = \varepsilon * s+b$. Thus, it would appear that, if coverage in some average (over μ) sense were satisfactory, the frequentist requirement for intervals to contain the true value at the requisite rate would be maintained. This, however, is not the generally accepted view by Particle Physicists, who would like not to undercover for any μ.

- Not too much overcoverage: Because coverage varies with μ, for methods that aim not to undercover anywhere, some overcoverage is inevitable. This corresponds to having some upper limits which are high, and this leads to undesirable loss of power in rejecting alternative hypotheses about the parameter's value.

- Short and empty intervals: These can be obtained for certain values of the observable, without resulting in undercoverage. They are generally regarded as undesirable for the reasons explained above.

- It is not obvious how to incorporate the above desiderata on interval length into an algorithm that would be useful for choosing between different methods for setting limits.

Two-Sided Intervals

An alternative to giving upper limits is to quote two-sided intervals. For example, a 68% confidence interval for the mass of the top quark might be 169 to 173 GeV/c^2, as opposed to its 90% upper limit being 174 GeV/c^2. Most of the difficulties and ambiguities mentioned above apply in this case too, together with some extra possibilities. Thus, while it is clear which of two possible upper limits is tighter, this is not necessarily so for two-sided intervals, where which is shorter may be metric dependent; the first of two intervals for a particle's lifetime τ may be shorter, but the second may be shorter when the ranges are quoted for decay rate ($= 1/\tau$). Also, there is more scope for choice of ordering rule for the frequentist Neyman construction, or for choosing the interval from the Bayesian posterior probability density.

It has been pointed out by Feldman and Cousins (1998) that an apparently innocuous procedure for choosing what result to quote may lead to undercoverage. Many physicists would quote an upper limit on any possible signal if their observation was not more than three standard deviations above the expected background, but a two-sided interval if their result was above this. With each type of interval constructed to give 90% coverage, there are some values of the parameter for which the coverage for this mixed procedure drops to 85%; Feldman and Cousins refer to this as "flip-flop." They circumvent the problem by using a "unified" approach, in which the method automatically yields upper limits for small values of the data, but two-sided intervals for larger measurements, while maintaining correct coverage for all possible true values of the signal.

Sensitivity

We have already mentioned the idea of quoting the sensitivity of a procedure, as well as the actual upper limit as derived from the observed data.8 For upper limits or for uncertainties on measurements, this can be defined as the median value that would be obtained if the procedure was repeated a large number of times. Using the median is preferable to the mean because (a) it is metric independent (i.e., the median lifetime upper limit would be the reciprocal of the median decay rate lower limit); and (b) it is much less sensitive to a few anomalously large upper limits or error estimates.

Punzi (2003) has drawn attention to the fact that this choice of definition for sensitivity has some undesirable features. Thus, minimizing the median upper limit for a search provides a different optimization from maximizing the median number of standard deviations for the significance of a discovery. Also, there is only a 50% chance of achieving the median result or better. Instead, for pre-defined levels α and CL, Punzi determines at what signal strength there is a probability of at least CL for establishing a discovery at a significance level α. This is what he quotes as the sensitivity, and is the signal strength at which we are sure to be able to claim a discovery or to exclude its existence. Below this, the presence or otherwise of a signal makes too little difference, and we may remain uncertain (see Figure 2).

CLs

This is a technique [Read (2000, 2004)] which is used for situations in which a discovery is not made, and instead various parameter values are excluded. For example, the Standard Model Higgs boson is such that, even before it is discovered, everything about it is well defined by theory except for its mass. The rate at which it is produced in a given experiment does depend on its mass. The failure to observe it can be converted into a mass range for the Higgs which is excluded (at some confidence level).

Figure 3 illustrates the expected distributions for some suitably chosen test statistic under two different hypotheses: the null H^0 in which there is only standard known physics, and H^1 which also includes some specific new particle, such as the Higgs boson. In the simplest case, the statistic could be simply the observed number of events n in some selected region.

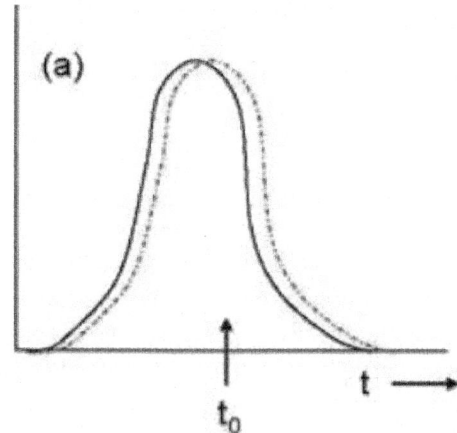

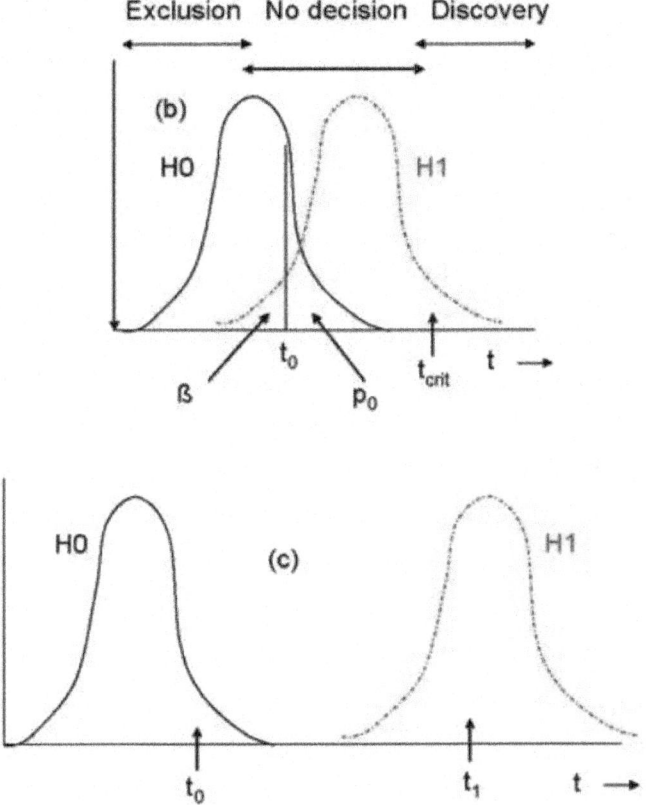

Figure 2: Punzi definition of sensitivity. Expected distributions for a statistic t (which in simple cases could be simply the observed number of events n), for H0 = background only (solid curves) and for H1 = background plus signal (dashed curves). In (a), the signal strength is very weak, and it is impossible to choose between H0 and H1. As shown in (b), which is for moderate signal strength, p0 is the probability according to H0 of t being equal to or larger than the observed t0. To claim a discovery, p0 should be smaller than some pre-set level α, usually taken to correspond to 5σ; tcrit is the minimum value of t for this to be so. Then the power function $1-\beta$ [equivalent to p1 in Figure 3(b)] is the probability according to the alternative hypothesis that t will exceed tcrit. According to Punzi, the sensitivity should be defined as the expected production strength of the signal such that $1-\beta$ exceeds another predefined level CL, for example, 95%. The exclusion region in (b) corresponds to t_0 in the 5% lower tail of H_1, while the discovery region has t0 in the 5σ upper tail of H0; there is a "No decision" region in between, as the signal strength in (b) is below the sensitivity value. The sensitivity is thus the signal strength above which there is a 95% chance of making a 5σ discovery. That is, the distributions for H_0 and H_1 are sufficiently separated that, apart possibly for the 5σ upper tail of H_0 and the 5% lower tail of H_1, they do not overlap. In (c) the signal strength is so large that there is no ambiguity in choosing between the hypotheses.

In Figure 3(c), the new particle is produced prolificly, and an experimental observation of n should fall in one peak or the other, and easily distinguish between the two hypotheses. In contrast, Figure 3(a) corresponds to very weak production of the new particle and it is almost impossible to know whether the new particle is being produced or not. The conventional method of claiming new particle production would be if n fell well above the main peak of the H_0 distribution; typically a p_0 value corresponding to 5σ would be required. In a similar way, new particle production would be excluded if n were below the main part of the H_1 distribution. Typically, a 95% exclusion region would be chosen (i.e., $1-p_1 \leq 0.05$). The CLs method aims to provide protection against a downward fluctuation of n in Figure 3(a), resulting in a claim of exclusion in a situation where the experiment has no sensitivity to the production of the new particle; this could happen in 5% of experiments.

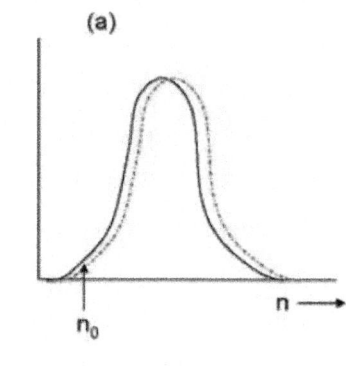

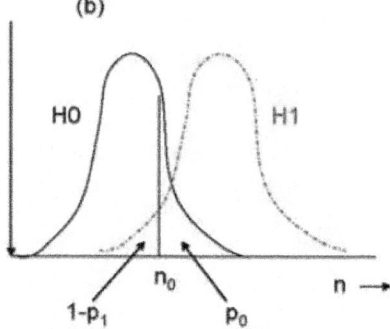

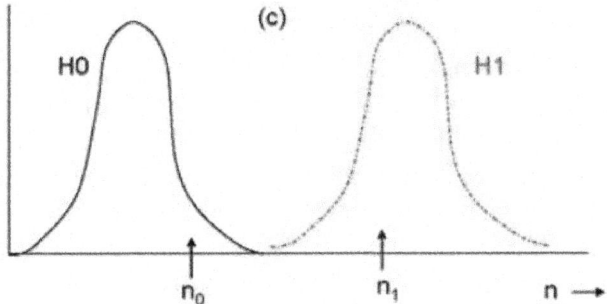

Figure 3: The CLs method. The expected distributions for a data statistic n are shown: (i) for the null hypothesis H_0 of background only (solid curve); and (ii) for H_1 (dashed curve), where there is also some exciting new physics, which tends to result in larger n. In (b), the tail areas of H_0 above the observed n_0 and of H_1 below n_0 are indicated by arrows; they correspond to probabilities p0 and $1 - p1$ respectively. Figure (c) shows a situation where the new physics is strongly produced, and H_0 and H_1 are well separated. Thus, n_0 would result in H_1 being excluded, while n_1 would be taken as evidence in favour of new physics. In (a), production is very weak, and the H_0 and H_1 curves are barely distinguishable. In order to protect against a downward fluctuation (statistic = n_0) in a situation like (a) resulting in an exclusion of H_1 when the curves are essentially identical, CLs is defined as $(1 - p_1)/(1 - p_0)$.

It achieves this by defining

$$CL_s = (1 - p_1)/(1 - p_0),$$

(1)

and requiring CLs to be below 0.05. From the definition, it is clear that CLs cannot be smaller than $1 - p_1$, and hence is a conservative version of the frequentist quantity $1 - p_1$. It tends to $1 - p_1$ when n lies above the H_0 distribution, and to unity when H_0 and H_1 are very similar. Statisticians may find CLs, which is the ratio of two p-values, to be lacking in formal justification. Its appeal to Particle Physicists is the protection it provides against excluding particles from data which have no sensitivity to them. We thus regard it as a conservative frequentist approach.

Nuisance Parameters

For calculating upper limits in the simple counting experiment described in Section 5, the nuisance parameters arise from the uncertainties in the background rate b and the acceptance ε. These uncertainties are usually quoted as σ_b and σ_ε (e.g., $b = 3.1 \pm 0.5$), and the question arises of what these errors mean. Sometimes they encapsulate the results of a subsidiary measurement, performed to estimate b or ε, and then they would express the width of the

Bayesian posterior or of the frequentist interval obtained for the nuisance parameters. However, in many situations, the errors may be based on a series of subsidiary measurements; they may involve Monte Carlo simulations, which have systematic uncertainties (e.g., related to how well the simulation describes the real data) as well as statistical errors; or they may reflect uncertainties or ambiguities in theoretical calculations required to derive b and/or ε. In the absence of further information the posterior is often assumed to be a Gaussian, usually truncated so as to exclude unphysical (e.g., negative) values. This may be at best only approximately true, and deviations are likely to be most serious in the tails of the distribution.

There are many methods for incorporating nuisance parameters in upper limit calculations. These include:

- Profile likelihood. The likelihood based on the data from the main and from the subsidiary measurements, is a function of the parameter of interest s and of the nuisance parameters. The profile likelihood $L_{prof}(s)$ is simply the full likelihood $L(s, b_{best}(s), \epsilon_{best}(s))$, evaluated at the values of the nuisance parameters that maximize the likelihood at each s. Then the profile likelihood is simply used to extract the limits on s, much as the ordinary likelihood could be used for the case when there are no nuisance parameters.

Rolke et al. (2005) have studied the behavior of the profile likelihood method for limits. Heinrich (2003a) had shown that the likelihood approach for estimating a Poisson parameter (in the absence of both background and of nuisance parameters) can have poor coverage at low values of the Poisson parameter. However, the profile likelihood seems to do better, probably because the nuisance parameters have the effect of smoothing away the fluctuating coverage observed by Heinrich.

- *Full Bayes.* When there is a subsidiary measurement, a prior is chosen for b (or ε), the data is used to extract the likelihood, and then Bayes' theorem is used to deduce the posterior for the nuisance parameter. This posterior from the subsidiary measurement is then used as the prior for the nuisance parameter in the main measurement (this prior could alternatively come from information other than a subsidiary experiment); together with the prior for s and the likelihood for the main measurement, the overall joint posterior for s and the nuisance parameter(s) is derived.10 This is then integrated over the nuisance parameter(s) to determine the posterior for s, from which an upper limit can be derived. Numerical examples of upper limits can be found in Heinrich et al. (2004), where the method is discussed in detail. Thus, for precisely determined backgrounds, the effect of a 10% uncertainty in ε can be seen for various measured values of n in Table 2. A plot of the coverage when the uncertainty in ε is 20% is reproduced in Figure 4.

It is not universally appreciated that the choice for the main measurement of a truncated Gaussian prior for ε and an (improper) constant prior for nonnegative s results in a posterior for s which diverges. Thus, numerical estimates of the relevant integrals are meaningless. Another problem comes from the difficulty of choosing sensible multi-dimensional priors. Heinrich has pointed out the problems that can arise for the above Poisson counting experiment, when it is extended to deal with several data channels simultaneously [Heinrich (2005)].

Fully frequentist: In principle, the fully frequentist approach to setting limits when provided with data from the main and from subsidiary measurements is straightforward: the Neyman construction is performed in the multidimensional space where the parameters are s and the nuisance parameters, and the data are from all the relevant measurements. Then the region in parameter space for which the observed data were likely is projected onto the s-axis, to obtain the confidence region for s. In practice, there are severe difficulties in writing a program to do this in a reasonable amount of time. To date, the largest number of parameters used is three [Nicola and Signorelli (2002)]. Another problem is that, unless a clever ordering rule is used for producing the acceptance region in data space for fixed values of the parameters, the projection phase leads to overcoverage, which can become larger as the number of nuisance parameters increases. Good ordering rules have been found for a version of the Poisson counting experiment [Punzi (2005)], and for the ratio of Poisson.

Table 2: 90% confidence level upper limits for the production rate s as a function of n, the observed number of events

n	$b = 0.0$	$b = 3.0$
0	2.35 (2.30)	2.35 (2.30)
3	6.87 (6.68)	4.46 (4.36)
6	10.88 (10.53)	7.80 (7.60)
9	14.71 (14.21)	11.56 (11.21)
20	28.27 (27.05)	25.05 (24.05)

The Poisson parameter μ = *s+b, where the expected background b is either 0.0 or 3.0, and is precisely known; and, whose true values is 1.0, is estimated in a subsidiary measurement with 10% accuracy. The numbers in brackets are the corresponding upper limits when ε is known precisely. At large n, the limits for b = 3.0 are 3 units lower than those for b = 0.0; the latter are approximately $n + 1.28\sqrt{n}$ at large n. The effect of the uncertainty in X is to increase the limits, and by a larger amount at large n. For n = 0, the Bayesian limits are independent of the expected background b.

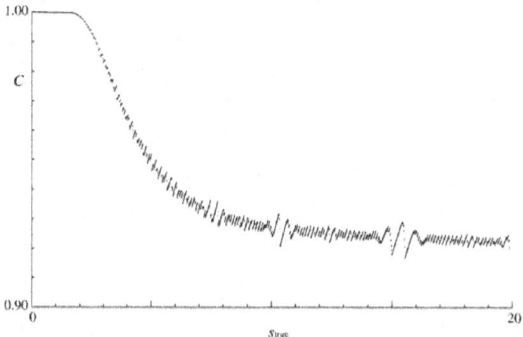

Figure. 4: The coverage C for the estimated 90% confidence level upper limit as a function of the true parameter strue. The background b = 3.0 is assumed to be known exactly, while the subsidiary measurement for □ gives a 20% accuracy. The discontinuities are a result of the discrete (integer) nature of the measurements. There appears to be no undercoverage.

means [Cousins (1998)], where the confidence intervals are tighter than those obtained by conditioning on the sum of the numbers of counts in the two observations. For the fully frequentist method, it is guaranteed that there will be no undercoverage for any combination of parameter true values. This is not so for any other method, and so most Particle Physicists would like assurance that the technique used does indeed provide reasonable coverage, at least for s. There is usually lively debate between frequentist and Bayesians as to whether coverage is desirable for all values of the nuisance parameter(s), or whether one should be happy with no or little undercoverage when experiments are averaged over the nuisance parameter true values. • Mixed. Because of the difficulty of performing a fully frequentist analysis in all but the simplest problems, an alternative approach [Cousins and Highland (1992)] is to use Bayesian averaging over the nuisance parameters, but then to employ a frequentist approach for s. The hope is that for most experiments setting upper limits, the statistical errors on the data are relatively large and so, provided the uncertainties in the nuisance parameters are not too large, the effect of the systematics on the upper limits will be small, and hence an approximate method of dealing with them may be justified. 5.5. Banff challenge. Given the large number of techniques available for extracting upper limits from data, especially in the presence of nuisance parameters, it was decided at the Banff meeting [Reid, Linnemann and Lyons (2006)] that it would be useful to compare the properties of the different approaches under comparable conditions. This led to the setting up of the "Banff Challege," which consisted of providing common data sets for anyone to calculate their upper limits. This was organized by Joel Heinrich, who reported on the performance of the various methods at the PHYSTAT-LHC meeting [Heinrich (2007)].

DISCOVERY ISSUES

Searches for new particles are an exciting endeavor, and will play an even bigger role with the start-up of the LHC at CERN, expected in 2008. The 2007 PHYSTAT Workshop at CERN [Prosper, Lyons and De Roeck (2007)] was devoted to statistical issues that arise in discoveryoriented analyses.

p-Values

In order to quantify the chance of the observed effect being due to an uninteresting statistical fluctuation, some statistic is chosen for the data. The simplest case would be the observed number n_0 of interesting events. Then the p-value is calculated, which is simply the probability that, given the expected background rate b from known sources, the observed number of events would fluctuate up to n_0 or larger. A small value of pindicates that the data are not very compatible with the theory (which may be because we do not understand our detector, rather than the theory being wrong). Particle physicists usually convert p into the number of standard deviations σ of a Gaussian distribution, beyond which the one-sided tail area corresponds to p. Thus, 5σ corresponds to a p-value of $3 * 10^{-7}$. This is done simply because it provides a number which is easier to remember, and not because Gaussians are relevant for every situation. Unfortunately, p-values are often misinterpreted as the probability of the theory being true, given the data. It sometimes helps colleagues clarify the difference between $p(A|B)$ and $p(B|A)$ by reminding them that the probability of being pregnant, given the fact that you are female, is considerably smaller than the probability of being female, given the fact that you are pregnant.

Nuisance Parameters

The calculation of p-values is complicated in practice by the existence of nuisance parameters. For example, for the simple situation described above, there could be some uncertainty in the estimated background. Although pivots are not generally used, there are numerous ways of incorporating nuisance parameters. These include:

- Conditioning: In simple cases with a single nuisance parameter, it may be possible to condition on the sum of the number of counts in the main and the subsidiary experiments, and then to use the binomial distribution to obtain the p-value.

- Plug-in p-value: The best estimate of the nuisance parameters is used to calculate p.

- Prior predictive p-value: The p-values are averaged over the nuisance parameters, weighted by their prior distributions.

- Posterior predictive p-value: This time, the posterior distributions of the nuisance parameters are used for weighting.

- Supremum p-value: The largest p-value for any possible value of the nuisance parameter is used. This is likely to be useful only when the nuisance parameter is forced to be within some range; or when there is only a finite number of possible alternative theoretical interpretations.

- Confidence interval: A confidence region of size $1 - \gamma$ is used for the nuisance parameter(s), and then the adjusted p-value is $p_{max} + \gamma$, where p_{max} is the largest p-value as the nuisance parameters are varied over their confidence region. Clearly, if it is desired to establish a discovery from p-values around 10^{-7} or smaller, then γ should be chosen at least an order of magnitude below this.

The properties of these and other methods are compared by Demortier (2007), while Cranmer (2007) has discussed some of them in the context of searches at the LHC, where the distributions in the tails of the probability distributions for data can be very relevant. Again, any experience of Statisticians about incorporating nuisance parameters could result in useful advice. The role of systematic effects is likely to be more serious here than for upper limits discussed in Section 5.4. This is because in upper limit situations the number of events is usually small, and so statistical errors dominate. In contrast, discovery claims have p-values of $3 * 10-7$ or smaller, and so tails of distributions are likely to be important.

Why 5σ?

Unfortunately the usually accepted ideal for claiming a discovery in Particle Physics is that p should correspond to at least 5σ. Statisticians almost invariably ask why we use such a stringent level. One answer is past experience: we have all too often seen interesting effects at the 3σ or 4σ level go away as more data are collected. Another is the multiple comparison problem, or "look elsewhere" effect. While the chance of obtaining a 5σ effect in one bin of a particular histogram is really small, it is to be remembered that histograms have many bins,11 they could be plotted with different selection criteria and different binning,12 and there are very many other histograms that were or could have been looked at in the course of the experiment.13 Thus, the chance of a 5σ fluctuation occurring somewhere in the data is much larger than might at first appear. Finally, physicists subconsciously incorporate Bayes' priors in assessing how likely they feel that they have discovered something new, and hence, whether they should claim a discovery. Thus, in deciding between the possibilities of a new discovery or of an undetected systematic effect, our priors might favor the latter, and hence, strong evidence for discovery is required from the data. It is not necessarily equitable to use a uniform standard for large generalpurpose experiments and for small ones with a specific aim; or for looking for a process which is expected, as compared with a very speculative search. But physicists and journal editors do like a defined rule rather than a flexible criterion, so this bolsters the 5σ standard. The general attitude is that, in the absence of a case for special pleading, 5σ is a reasonable requirement. In any case, it is largely a semantic issue, in that physicists finding a 4.5σ effect would clearly report it, using judiciously chosen wording to describe the status of their observation. Statisticians also ask whether we really believe our models out into the extreme tails of the distributions. In general, this may be so—counting experiments are expected to follow Poisson distributions, with small corrections for possible long time-scale drifts in detector calibrations; and particle decays usually are described by exponential distributions in time. However, the situation is much less clear for nuisance parameters, where error estimates may be less rigorous, and their distribution is often assumed to be Gaussian (or truncated Gaussian) by default. The effect of these uncertainties on very small p-values needs to be investigated case-by-case. We also have to remember that p-values merely test the null hypothesis. A more sensitive way to look for new physics is via the likelihood ratio or the differences in χ^2 for the two hypotheses, that is, with and without the new effect. Thus, a very small p-value on its own is usually not enough to make a convincing case for discovery.

Repetitions in Time

A typical experiment at a large accelerator may collect data over 10–15 years. The same search for a new effect will typically be repeated once or twice each year as more data is collected. Does this constitute another factor of ≈20 in the number of opportunities for a statistical fluctuation to appear? Our reply is "No." If there had been a 6σ signal with half the data (which resulted in a claim for discovery), which then became only 3σ with more data, this would be grounds for downplaying the earlier discovery claim. Thus, at any time, there is only one set of data (everything) that is relevant.

Combining p-Values

In looking for a given new effect, there may be several separate and uncorrelated analyses which are relevant. These could correspond to different decay possibilities for the new particle or different experiments looking for the same signal. Thus, if the p-values for the null hypothesis (i.e., no new physics) for the separate analyses were 10^{-6} and 0.1, what is the corresponding p-value for the pair of results14? The unambiguous answer is that there is no unique recipe for combining them [CDF (2007), Cousins (2007)]. There is no single way of taking a uniform distribution in two variables, and finding a transformation pcomb(p1, p2) that converts it into a uniform distribution of the single variable pcomb.

Two popular recipes involve asking what is the probability that the smaller p-value will be 10^{-6} or smaller or that the product is below $p1 * p2 = 10^{-7}$. None of the possible methods has the property that in combining 3 p-values, the same answer is obtained if p1 is first combined with p2, and then the result is combined with p3; or whether some different ordering is used. Clearly, it is important to decide what combination method should be used, without reference to the specific data.

Peak above Smooth Background

When comparing two hypotheses with our data, we can use the numerical values of the two χ^2 quantities. For example, we may be fitting a smooth distribution by a power series, and wonder whether we need a quadratic term, or whether a linear expression would suffice. Alternatively, we may want to assess whether a mass spectrum favors the existence of a peak on top of a smooth background, as compared with just the smooth background. Qualitatively, if the extra term(s) are unnecessary, they will result in a relatively small reduction in χ^2, while if they really are required, the reduction could be larger. It is sometimes possible to be quantitative about the expected reduction when the extra terms are not needed [Wilks (1938)]. If we are in the asymptotic regime, and if the hypotheses are nested, and if the extra parameters of the larger hypothesis are defined under the smaller one, and in that case do not lie on the boundary of their allowed region, then the difference in χ^2 should itself be distributed as a χ^2, with the number of degrees of freedom equal to the number of extra parameters. An example that satisfies this is provided by the different order polynomials. Provided we have a large amount of data, we expect the difference in χ^2 to have one degree of freedom, so a value larger than around 5 would be unlikely.

A contrast is provided by a smooth background $C(x)$ compared with a background plus peak, $C(x) + A\exp[-0.5 * (x - x0)^2/\sigma^2]$. The extra parameters for the peak are its amplitude, position and width: A, x0 and σ respectively. Again, the hypotheses are nested, in that $C(x)$ is just a special case of the peak plus background, with A = 0. However, although A is de- fined in the background only case, x0 and σ are not, as their values become completely irrelevant when A = 0. Furthermore, unless the peak plus background fit allows A to be negative, zero is on the boundary of its allowed region. We thus should not expect the difference of the χ^2 quantities itself to be distributed as a χ^2 [Protassov et al. (2002), Demortier (2006)]. To assess the significance of a particular χ^2 difference, this unfortunately means that we have to obtain its distribution ourselves, presumably by Monte Carlo. If we want to find out probabilities of statistical fluctuations at the 10^{-6} level, this requires a lot of simulation, and probably needs us to use something better than brute force.

Another example of comparing hypotheses by their χ^2 values is given in Section 11.3. The problem of nonstandard limiting distributions for χ^2 tests has a substantial statistical literature [see, e.g., Self and Liang (1987) and Drton (2007)].

GOODNESS-OF-FIT

With sparse data, the unbinned likelihood method is a good one for estimating parameters of a model. In order to understand whether these estimates of the parameters are meaningful, we need to know whether the model provides an adequate description of the data. Unfortunately, as emphasised by Heinrich (2003b), maximum likelihood is often insensitive to whether or not the data agree with the model. It would be very useful to have a way of utilizing the unbinned likelihood so that it does provide a measure of the goodness-of-fit. The standard method loved by Particle Physicists is χ^2 . This, however, is only applicable to binned data (i.e., in a one or more dimensional histogram). Furthermore, it loses its attractive feature that its distribution is modelindependent when there are not enough data, which is likely to be so in the multi-dimensional case.

An alternative that is used for sparse one-dimensional data is the Kolmogorov– Smirnov (KS) approach or one of its variants. However, in the presence of fitted parameters, simulation is again required to determine the expected distribution of the KS-distance. Also because of the problem of how to order the data, it is not used by Particle Physicists in multi-dimensional situations. Aslan and Zech (2004, 2005) have described a method that can be used with sparse multi-dimensional data.15 It compares two separate sets of events, which could be data and simulation based on a theoretical model or two sets of data taken under slightly different conditions, etc. The first set of points are assigned positive electric charges, and the second set negative ones, and then the "electrostatic energy" of the system is calculated as $E = \Sigma\Sigma q_i * q_j * f(d_{ij})$, where the summation extends over all pairs of observations; qi is the charge of the ith observation; and $f(d_{ij})$ is a function of the distance d_{ij} between observations i and j. For real electrostatics in 3 dimensions, f(d) is proportional to 1/d, but here it can be chosen to give desirable behavior; Aslan and Zech favor $-\ln(d +)$, where is a small constant to avoid problems as d tends to zero. This method requires the choice of a metric for each of the observables, and it also needs simulation to determine the expected distribution of E assuming the two distributions are identical. Aslan and Zech find that their method compares favorably with other approaches (e.g, χ^2, KS and its variants, etc.) in rejecting alternative hypotheses in various one-dimensional problems.

BLIND ANALYSES

These are becoming increasingly popular in Particle Physics, as a means of avoiding personal bias affecting the result. They involve keeping part of the data unseen by the physicists, until the data selection procedure and the analysis method have been completely defined, all correction procedures specified, etc. The original suggestion to use a blind analysis for a Particle Physics experiment was due to Luis Alvarez. An experiment at Stanford had looked for quarks, by measuring the residual charge on small spheres that were levitated in a superconducting magnet. If a single free quark was present in a sphere, the residual charge would be a third or two-thirds of the electron's charge. Several of the balls tested indeed yielded such values. A potential problem was that large corrections had to be applied to the raw data in order to extract the final result for the charge. The suspicion was that maybe the experimenters were (subconsciously) applying corrections until the value turned out to be "satisfactory." The blind approach would involve the computer adding a random number to the raw value of the charge, which would then be corrected until the experimentalists were satisfied, and only then would the computer subtract the random number to reveal the final answer for that sphere.16 There are various methods of performing blind analyses [Klein and Roodman (2005)], most of which aim to allow the experimentalists to look at some of the real data, in order to perform checks that nothing is terribly wrong. Some of these are as follows:

- The computer adds a random number to the data, which is only subtracted after all corrections are applied. This was the method suggested by Alvarez.
- Use only Monte Carlo to define the procedure. This completely avoids the danger of allowing the data to determine the procedure to be used, but suffers from the drawback that the data cannot be compared with the Monte Carlo, to check that the latter is reasonable.
- Use only a fraction of the data for defining the procedure. Then this is

held fixed for the remainder of the data. In principle, an optimization can be employed to determine the fraction to be kept open, but, in practice, this is often decided by choosing a semi-arbitrary time after which the future data is kept blind.

- The signal region is defined as a certain part of multi-dimensional space, and this is kept hidden, but all other regions, including those adjacent to the signal, are available for inspection.

- Keep the Monte Carlo parameters hidden. This is a technique used by the TWIST experiment in their high statistics precision determination of parameters associated with muon decay. The procedure involves comparing the data with various simulated sets, generated with a series of different parameter values. The data and the simulations are both visible, but the parameter values used to generate the simulations are kept hidden.

- Keep visible only a fraction of the contents of each bin of a histogram. This is used by the MINOS experiment searching for neutrino oscillations; these would affect the energy distribution of the observed events. By keeping visible different unknown fractions of the data in each bin, the energy spectral shape cannot be determined from the visible part of the data.

If several different groups within the same collaboration are performing similar analyses for extracting some specific parameter, then it is desirable to fix the procedure for selecting which result to present, or alternatively how to combine the separate results. This should be done before the results are seen, and is worth doing even if the individual analyses were not "blind." A question that arises with blind analyses is whether it should be permitted to modify the analysis after the data had been unblinded. It is generally agreed that this should not be done. . . unless everyone would regard it as ridiculous not to do so. For example, if a search for rare events yielded 10 candidates over the course of a year's run, and it was found that all of these occurred on Sunday mornings at precisely 1:17 a.m., it would be prudent to do some further investigation before publishing. If "post-unblinding" modification of the procedure is performed, this should be made clear in any publication.

COMBINING RESULTS

A commonly used procedure is to combine N different uncorrelated measurements $a_i \pm \sigma_i$ of the same physical quantity a. When the measurements are believed to be Gaussian distributed about the true value a_{true}, the well-known result is that the best estimate $a_{best} \pm \sigma_{best}$ is given by

$$a_{\text{best}} = \Sigma a_i * w_i / \Sigma w_i, \qquad \sigma_{\text{best}} = 1/\sqrt{\Sigma w_i}, \tag{2}$$

where the weights are defined as $w_i = 1/\sigma_i^2$ This is readily derived from minimizing with respect to a a weighted sum of squared deviations

$$S(a) = \Sigma (a_i - a)^2 / \sigma_i^2. \tag{3}$$

The extension to the case where the individual measurements are correlated (as is often the case for analyses using different techniques on the same data) is straightforward: S becomes $\Sigma\Sigma(a_i - a) * H_{ij} * (a_j - a)$, where H is the inverse error matrix. There are, however, practical details that complicate its application. For example, in the above formula, the σ_i are supposed to be the true accuracies of the measurements. Often, all that we have available are estimates of their values. Problems arise in situations where the error estimate depends on the measured value a_i. For example, in counting experiments with Poisson statistics, it is typical to set the error as the square root of the observed number. Then a downward fluctuation in the observation results in an overestimated weight, and abest is biassed downward. If instead the error is estimated as the square root of the expected number a, the combined result is biassed upward—the increased error reduces S at larger a. A way round this difficulty has been suggested by Lyons, Martin and Saxon (1990). Another problem arises when the individual measurements are very correlated. When the correlation coefficient of two uncertainties is larger than σ_1/σ_2 (where σ_1 is the smaller error), a_{best} lies outside the range of the two measurements. As the correlation coefficient tends to +1, the extrapolation becomes larger, and is very sensitive to the exact value assumed for the correlation coefficient. The situation is aggravated by the fact that σ_{best} tends to zero. This is usually dealt with by selecting one of the two analyses, rather than trying to combine them.

Another extension of this procedure is for combining N pairs of correlated measurements (e.g., the gradient and intercept of a straight line fit to several sets of data). The prescription to be adopted for scaling the errors when the individual measurements are somewhat discrepant has complications.

ACCURACY OF ANSWER

Sometimes a result appears to be more accurate than is justified. This can arise when an upper limit is much lower than the sensitivity of the procedure (e.g., when the observed number of events in a counting experiment is smaller than the expected background) or when by chance individual observations happen to lie close to each other. This can cause problems in deciding which measurement is "better." This can be relevant in choosing which of several competing analyses on the same data to quote as the result of the experiment; or in combining different results (see previous section). In the former situation, if the estimated error increases with the estimated value, choosing the result with the smallest estimated error can produce a downward bias. On the other hand, using the smallest expected error can cause us to ignore an analysis which had a particularly favorable statistical fluctuation, which produced a result that was genuinely more precise than expected.18 How to deal with this situation in general is an open question. It has features in common with the problem of measuring a voltage by choosing at random a voltmeter from a cupboard containing meters of different sensitivities [Cox (1958)].

RECENT IMPROVEMENTS IN UNDERSTANDING

In this section we list a few of the issues on which Particle Physicists have recently improved their understanding of statistical issues. To those can be added a few already discussed above (see Section 6.6 and the remarks about unbinned likelihoods in the first paragraph of Section 7).

Number of Degrees of Freedom

If we construct the weighted sum of squares S between a predicted theoretical curve and some data in the form of a histogram, provided the Poisson distribution of the data can be approximated by a Gaussian (and the theory is correct, the data are unbiassed, the error estimates are correct, etc.), asymptotically19 S will be distributed as χ^2 with the number of degrees of freedom $v = n - f$, where n is the number of data points and f is the number of free parameters whose values are determined in the fit.

The relevance of the asymptotic requirement can be seen by imagining fitting a more or less flat distribution by the expression $N(1 + 10-6 \cos(x - x_0))$, where the free parameters are the normalization N and the phase x_0. It is clear that, although x0 is left free in the fit, because of the 10^{-6} factor, it will have a negligible effect on the fitted curve, and hence will not result in the typical reduction in S associated with having an extra free parameter. Of course, with an enormous amount of data, we would have sensitivity to x_0, and so asymptotically it does reduce v by one unit, but not for smaller amounts of data.

Another example involves the search for neutrino oscillations. The neutrino energy spectrum is fitted by a survival probability P of the form

$$P = 1 - A\sin^2(C * \Delta m^2), \tag{4}$$

where C is a known function of the neutrino energy and the length of its flight path, A is a parameter which depends on the neutrino mixing angle, and Δm^2 is the difference in mass squared of the relevant neutrino species. For small values of $C * \Delta m^2$,

$$P \approx 1 - A(C * \Delta m^2)^2. \tag{5}$$

Thus, the survival probability depends only on the two parameters in the combination $A (\Delta m^2)^2$. Because this combination is all that we can hope to determine, we effectively have only one free parameter rather than two. Of course, an enormous amount of data can manage to distinguish between $\sin(C * \Delta m^2)$ and $C * \Delta m^2$, and so asymptotically we have two free parameters as expected.

It would be useful to have some indication of when data are near enough to asymptopia, so as to avoid the necessity for Monte Carlo calculations of the expected distribution of S.

$\Delta(\ln L) = 0.5$ Rule.

In the maximum likelihood approach to parameter determination, the best value λ_0 of a parameter is determined by finding where the likelihood maximizes; and its error σ_λ is estimated by finding how much the parameter must be changed in order for the logarithm of the likelihood to decrease by 0.5 as compared with the maximum.20 From a frequentist viewpoint, this should ideally result in the range from $\lambda_0 - \sigma_\lambda$ to $\lambda_0 + \sigma_\lambda$ having 68% coverage.

If the measurement is distributed about the true value as a Gaussian with a constant width, then exact coverage is obtained, but in general this is not so. For example, Heinrich (2003a) has investigated the properties of the likelihood approach to estimate μ, the mean of a Poisson, when n_{obs} events are observed. Because n_{obs} is a discrete variable, the coverage is a discontinuous function of μ, and varies from 100% at $\mu = 0$ down to 30% at $\mu \approx 0.5$.21

Two Hypotheses via χ^2

Assume we have a histogram with 100 bins, and that we are using a χ^2 method for fitting it with a function with one free parameter. We expect to obtain a χ^2 value of 99 ± 14. Thus, if p_0, the best value of the parameter, yields a χ^2 of 85, we would regard that as very satisfactory. However, a theoretical colleague has a model which predicts that the parameter should have a different value p_1, and wants to know what the data has to say about that. We test this by calculating the χ^2 for that p_1 and obtain a value of 110. We appear to have two contradictory conclusions:

- p1 is satisfactory: This is based on the fact that the relevant χ^2 of 110 is well within the expected range of 99 ± 14.

- p1 is ruled out: The uncertainty on p is estimated by seeing how much it must change from its optimum value in order to make χ^2 increase by 1 unit. For this data, $\chi^2 (p_1)$ is 25 units larger than $\chi^2 (p_0)$, and so, assuming that the behavior of χ^2 in the neighborhood of the minimum is parabolic, p_1 is ruled out at the 5 standard deviation level.

Unfortunately, many physicists, over-impressed by the fact that $\chi^2 (p1)$ appears to be satisfactory, are reluctant to accept that p0 is strongly favored by the data. A similar argument applies to comparing a given set of data with 2 separate hypotheses, for example, fitting a histogram with an exponential or a straight line. Again the difference between the χ^2 quantities provides better discrimination between the hypotheses than do the individual χ^2 [Lyons (1999)]. There are of course other ways of comparing two hypotheses e.g. likelihood ratio, Bayes factor, Bayesian information criterion, etc. Trotta (2008) has discussed their application in cosmology.

CONCLUSIONS

It is clear that there are many practical issues to be resolved in Particle Physics. Some of these may be of interest to Statisticians. With analyses becoming more and more complex, we would welcome more active involvement that would lead to improved analyses of our data. Any suggestions regarding improvements in the approaches outlined in this review would also be appreciated.

ACKNOWLEDGMENTS

I wish to acknowledge the patience and expertise of David Cox, Brad Efron and Michael Stein and also of other Statisticians too numerous to list, in explaining statistical issues to me; the ones who have contributed to the PHYSTAT meetings have been particularly helpful. My understanding of the practical application of statistical techniques has improved considerably as a result of discussions with many experimental Particle Physics colleagues. I especially want to thank the members of the CDF Statistics Committee and Bob Cousins. To all of you, I am most grateful.

REFERENCES

1. Aslan, B. and Zech, G. (2004). A multivariate two-sample test based on the concept of minimum energy. J. Stat. Comp. Simul. 75 109. MR2117010
2. Aslan, B. and Zech, G. (2005). Statistical energy as a tool for binning-free multivariate goodness of fit tests, two-sample comparison and unfolding. Nuclear Instruments and Methods A537 626.

3.   Belikov, J. (2007). ALICE statistical wish-list. Available at http://phystat-lhc.web.cern.ch/phystat-lhc/.

4.   CDF Statistics Committee (2007). Frequently asked questions. Available at http://www-cdf.fnal.gov/physics/statistics/statistics faq.html#iptn4.

5.   Cheung, H. and Lyons, L. (2000). FNAL confidence limits workshop. Available at http://conferences.fnal.gov/CLW/.

6.   Cousins, R. (1998). Improved central confidence intervals for the ratio of Poisson means. Nuclear Instruments and Methods A417 391–399.

7.   Cousins, R. (2007). Annotated bibliography on some papers on combining significances or p-values. Available at arXiv:0705.2209.

8.   Cousins, R. D. and Highland, V. L. (1992). Incorporating systematic uncertainties into an upper limit. Nuclear Instruments and Methods A320 331.

9.   Cox, D. R. (1958). Some problems connected with statistical inference. Ann. Math. Statist. 29 357–372. MR0094890

10.   Cranmer, K. (2007). Progress, challenges and future of statistics at the LHC. Available at http://phystat-lhc.web.cern.ch/phystat-lhc/2008-001.pdf.

11.   Cuadras, C. M., Fortiana, J. and Oliva, F. (1997). The proximity of an individual to a population with applications to discriminant analysis. J. Classification 14 117–136.MR1449744

12.   Cuadras, C. M. and Fortiana, J. (2003). Distance-based multivariate two sample tests. Available at http://www.imub.ub.es/publications/preprints/pdf/ Cuadras-Fortiana.334.pdf. MR2091618

13.   Demortier, L. (2005). Bayesian reference analysis. Available at http://www.physics.ox.ac.uk/phystat05/proceedings/files/demortier-refana.ps. MR2270218

14.   Demortier, L. (2006). Setting the scene for p-values. Available at www.birs.ca/workshops/2006/06w5054/report06w5054.pdf.

15.   Demortier, L. (2007). P-values and nuisance parameters. Available at http://phystat-lhc.web.cern.ch/phystat-lhc/2008-001.pdf.

16.   Drton, M. (2007). Likelihood ratio tests and singularities. Available at http://front.math.ucdavis.edu/0703.5360.

17.   Feldman, G. J. and Cousins, R. D. (1998). Unified approach to the classical statistical analysis of small signals. Phys. Rev. D 57 3873–3889.

18.   Friedman, J. H. (2003). Recent advances in predictive (machine) learning. Available at http://www-stat.stanford.edu/~jhf/ftp/machine.pdf.

19. Friedman, J. H. (2005). Separating signal from background using ensembles of rules. Available at http://www.physics.ox.ac.uk/phystat05/proceedings/files/friedman phystat.pdf.

20. Gross, E. (2007). ATLAS and CMS statistical wish-list. Available at http://phystat-lhc.web.cern.ch/phystat-lhc/2008-001.pdf.

21. Heinrich, J. (2003a). Coverage of error bars for Poisson data. Available at http://www-cdf.fnal.gov/publications/cdf6438 coverage.pdf.

22. Heinrich, J. (2003b). Pitfalls of goodness-of-fit from likelihood. Available at http://www.slac.stanford.edu/econf/C030908/papers/MOCT001.pdf.

23. Heinrich, J. (2005). The Bayesian approach to setting limits: What to avoid. Available at http://www.physics.ox.ac.uk/phystat05/proceedings/files/heinrich.ps.

24. Heinrich, J. (2007). Review of Banff challenge on upper limits. Available at http://phystat-lhc.web.cern.ch/phystat-lhc/2008-001.pdf.

25. Heinrich, J. and Lyons, L. (2007). Systematic errors. Annual Reviews of Nuclear and Particle Science 57 145–169.

26. Heinrich, J. et al. (2004). Interval estimation in the presence of nuisance parameters. 1. Bayesian approach. CDF Note 7117. Available at http://www-cdf.fnal.gov/publications/cdf7117 bayesianlimit.pdf.

27. Hocker, A. et al. ¨ (2007). TMVA—Toolkit for multivariate data analysis. Available at http://phystat-lhc.web.cern.ch/phystat-lhc/2008-001.pdf.

28. James, F., Lyons, L. and Perrin, Y. (2000). Workshop on confidence limits. CERN Yellow Report 2000-05. http://documents.cern.ch/cgi-bin/setlink?base= cernrep&categ=Yellow Report&id=2000-005#top.

29. Klein, J. R. and Roodman, A. (2005). Blind analysis in nuclear and particle physics.Annual Review of Nuclear and Particle Physics 55 141.

30. Linnemann, J. (2007). A pitfall in evaluating systematic errors. Available at http://phystat-lhc.web.cern.ch/phystat-lhc/2008-001.pdf.

31. Lyons, L. (1999). Comparing two hypotheses. Available at http://www-cdf.fnal.gov/physics/statistics/statistics recommendations.html.

32. Lyons, L. (2008). Supplement to "Open statistical issues in particle physics." DOI: 10.1214/08-AOAS163SUPP.

33. Lyons, L., Martin, A. and Saxon, D. (1990). On the determination of the B lifetime by combining the results of different experiments. Phy. Rev. D 41 982.

34. Lyons, L., Mount, R. and Reitmeyer, R., eds. (2003). PHYSTAT 20003.

35. eConf C030908, SLAC-R-703. Available at http://www.slac.stanford.

edu/econf/ C030908/proceedings.html.

36. Lyons, L. and Unel, M. K. " (2005). Statistical problems in particle physics, astrophysics and cosmology. Imperial College Press, London. Available at http://www.physics.ox.ac.uk/phystat05/. MR2270215

37. Narsky, I. (2000). Comparison of upper limits. Available at http:// conferences.fnal.gov/cl2k/copies/inarsky.pdf.

38. Narsky, I. (2006). StatPatternRecognition: A C++ package for multi-variate classification. Available at http://www.hep.caltech.edu/~narsky/ SPR/SPR narsky chep2006.pdf.

39. Neal, R. (2007). Computing likelihood functions when distributions are defined by simulators with nuisance parameters. Available at http:// phystat-lhc.web.cern.ch/phystat-lhc/2008-001.pdf.

40. Nicolo, D. and Signorelli, G. (2002). Application of strong confidence to the CHOOZ experiment with frequentist inclusion of nuisance parameters. Available at http://www.ippp.dur.ac.uk/old/Workshops/02/ statistics/proceedings/signorelli. pdf.

41. Particle Data Group (2006). J. Phys. G: Nucl. Part. Phys. 33 1 (see page 14).

42. Prosper, H. B. (2002). Multivariate analysis: A unified perspective. Available at www.ippp.dur.ac.uk/Workshops/02/statistics/papers/prosper HBPDurham2002.ppt.

43. Prosper, H. B., Lyons, L. and De Roeck, A. (2007). PHYSTAT-LHC Workshop at CERN on Statistical Issues for LHC Physics. Available at http://phystat-lhc.web.cern.ch/phystat-lhc/2008-001.pdf.

44. Protassov, R. et al. (2002). Statistics: Handle with care. Detecting multiple model components with the likelihood ratio test. Astrophysics J. 571 545–559.

45. Punzi, G. (2003). Sensitivity of searches for new signals and its optimization. Available at http://www.slac.stanford.edu/econf/C030908/ papers/MODT002.pdf.

46. Punzi, G. (2005). Ordering algorithms and confidence intervals in the presence of nuisance parameters. Available at http://www.physics.ox.ac. uk/ phystat05/proceedings/files/Punzi PHYSTAT05 final.pdf.

47. Read, A. L. (2000). Modified frequentist analysis of search results (the CLs method). Available at http://doc.cern.ch/archive/electronic/cern/ preprints/open/ open-2000-205.pdf.

48. Read, A. L. (2004). Presentation of search results—the CLs method. J. Phys. G: Nucl. Part. Phys. 28 2693–2704.

49. Reid, N. (2007). Some aspects of design of experiments. Available at http://phystat-lhc.web.cern.ch/phystat-lhc/2008-001.pdf.

50. Reid, N., Linnemann, J. and Lyons, L. (2006). Workshop on statistical inference problems in high energy physics and astronomy. Available at http://www.birs.ca/birspages.php?task=displayevent\&event id=06w5054.

51. Roe, B. P. (2007). Statistical errors in Monte Carlo estimates of systematic errors. Nuclear Instruments and Methods A570 159–164.

52. Rolke, W. A., Lopez, A. M. and Conrad, J. (2005). Limits and confidence intervals in the presence of nuisance parameters. Nuclear Instruments and Methods A551 493–503.

53. Self, S. G. and Liang, K. Y. (1987). Asymptotic properties of maximum likelihood estimators and likelihood ratio test under non-standard conditions. J. Amer. Statist. Assoc. 82 605–610. MR0898365

54. Sen, B., Walker, M. and Woodroofe, M. (2008). On the unified method with nuisance parameters. Statist. Sinica. To appear.

55. Trotta, R. (2008). Bayes in the sky: Bayesian inference and model selection in cosmology. Contemporary Physics 49 71–104.

56. Whalley, M. and Lyons, L., eds. (2002). Advanced statistical techniques in particle physics, Durham IPPP/02/39. Available at

57. http://www.ippp.dur.ac.uk/old/Workshops/02/statistics/proceedings. shtml. Wilks, S. S. (1938). The large-sample distribution of the likelihood ratio for testing composite hypotheses. Ann. Math. Statist. 9 60–62.

58. Xie, Y. (2007). LHCb statistical wishlist. Available at http://phystat-lhc. web.cern.ch/phystat-lhc/2008-001.pdf.

CITATION

CHAPTER 1

Denis Perret-Gallix, Computational particle physics for event generators and data analysis, Doi: 10.1088/1742-6596/454/1/012051.

CHAPTER 2

Hsuan Tung Peng and Yew Kam Ho, Statistical Correlations of the N-particle Moshinsky Model, doi: 10.3390/e17041882

CHAPTER 3

Giertz, H. (2015) A Novel Elementary Particle Theory Based on External Energy Absorbed and Re-Emitted by Atoms. Journal of Modern Physics, 6, 157-167. doi: 10.4236/jmp.2015.62021.

CHAPTER 4

Ajaltouni, Z. (2014) Symmetry and relativity: From classical mechanics to modern particle physics. Natural Science, 6, 191-197. doi: 10.4236/ns.2014.64023.

CHAPTER 5

Edwin Zong, The Origin of Particles, Journal of Nuclear and Particle Physics, Vol. 5 No. 3, 2015, pp. 45-51. doi: 10.5923/j.jnpp.20150503.01.

CHAPTER 6

Edwin Zong , The Unification of Particle and Energy, Journal of Nuclear and Particle Physics, Vol. 5 No. 4, 2015, pp. 79-83. doi: 10.5923/j.jnpp.20150504.02.

CHAPTER 7

H. M. M. Mansour, Single Particle Potentials and Three-Body Forces, Journal of Nuclear and Particle Physics, Vol. 4 No. 6, 2014, pp. 171-175. doi: 10.5923/j.jnpp.20140406.03.

CHAPTER 8

Edwin Zong, The Real Universe, Journal of Nuclear and Particle Physics, Vol. 5 No. 2, 2015, pp. 21-29. doi: 10.5923/j.jnpp.20150502.01.

CHAPTER 9

H. Mansour, Hend M. Saad, M. Aziz, Analysis of Reactivity - Initiated Accident for Control Rods Ejection, Journal of Nuclear and Particle Physics, Vol. 3 No. 4, 2013, pp. 45-54. doi: 10.5923/j.jnpp.20130304.01.

CHAPTER 10

M. M. Ahmed , S. K. Das , M. A. Haydar , M. M. H. Bhuiyan , M. I. Ali , D. Paul , Study of Natural Radioactivity and Radiological Hazard of Sand, Sediment, and Soil Samples from Inani Beach, Cox's Bazar, Bangladesh, Journal of Nuclear and Particle Physics, Vol. 4 No. 2, 2014, pp. 69-78. doi: 10.5923/j.jnpp.20140402.04.

CHAPTER 11

Mohamed Y. Abou-zeid, Seham S. El-zahrani, Hesham M. Mansour, Mathematical Modeling for Pulsatile Flow of a Non-Newtonian Fluid with Heat and Mass Transfer in a Porous Medium between two Permeable Parallel Plates, Journal of Nuclear and Particle Physics, Vol. 4 No. 3, 2014, pp. 100-115. doi: 10.5923/j.jnpp.20140403.03.

CHAPTER 12

B. M. R. Faisal, M. A. Haydar, M. I. Ali, D. Paul, R. K. Majumder, M. J. Uddin, Assessment of Natural Radioactivity and Associated Radiation Hazards in Topsoil of Savar Industrial Area, Dhaka, Bangladesh,Journal of Nuclear and Particle Physics, Vol. 4 No. 4, 2014, pp. 129-136. doi: 10.5923/j.jnpp.20140404.03.

CHAPTER 13

H. M. Mansour, A. A. I. Khalil, M. Y. Helali, M. Mansour, Development of the Stability on the Laser System Used at Satellite Laser Ranging Station, Journal of Nuclear and Particle Physics, Vol. 4 No. 1, 2014, pp. 7-16. doi: 10.5923/j.jnpp.20140401.02.

CHAPTER 14

Louis Lyons, Open Statistical Issues in Particle Physics, DOI: 10.1214/08-AOAS163

INDEX